# Wheelchair Housing Design Guide

Third edition

**Habinteg Housing Association**
Holyer House
20–21 Red Lion Court
London EC4A 3EB
Tel. 020 7822 8700
www.habinteg.org.uk

**Royal College of Occupational Therapists Specialist Section – Housing**
106–114 Borough High Street
London SE1 1LB
Tel. 020 7357 6480
www.rcot.co.uk/aboutus/specialist-sections/housing-rcots-ss

**Authors**
Michelle Horn, BA (Hons) Architecture
Centre for Accessible Environments at Habinteg

Jacqui Smith, BScOT
Centre for Accessible Environments at Habinteg

Jacquel Runnalls, DipCOT, MSc
Accessibility & Inclusive Design
RCOT Specialist Section – Housing

Kate Sheehan, DipCOT
RCOT Specialist Section – Housing

We are also grateful to Adam Thomas of Adam Thomas Consulting for co-authoring Chapter 8: Using the kitchen.

© Habinteg Housing Association, 2018
Third Edition

Published by RIBA Publishing,
66 Portland Place, London, W1B 1NT

ISBN 978-1-85946-828-9 /
978-1-85946-829-6 (PDF)

The right of Habinteg Housing Association to be identified as the Author of this Work has been asserted in accordance with the Copyright, Designs and Patents Act 1988 sections 77 and 78.

All rights reserved. No part of this publication may be reproduced, stored in a retrieval system, or transmitted, in any form or by any means, electronic, mechanical, photocopying, recording or otherwise, without prior permission of the copyright owner.

British Library Cataloguing-in-Publication Data
A catalogue record for this book is available from the British Library.

Commissioning Editor: Ginny Mills
Project Manager: Alasdair Deas
Designed and typeset by CSM Live
Printed and bound by Page Bros, Norwich
Cover image: Alison Grant

While every effort has been made to check the accuracy and quality of the information given in this publication, neither the Author nor the Publisher accept any responsibility for the subsequent use of this information, for any errors or omissions that it may contain, or for any misunderstandings arising from it.

www.ribapublishing.com

---

This guide is based on the 2015 edition (incorporating the 2016 amendments) of *Approved Document M Volume 1: Dwellings, M4(3) Category 3: Wheelchair user dwellings*. Any subsequent amendments to the Approved Document or publication of supplementary guidance should be taken into account when using this guide. While every effort has been made to ensure that all details are accurate, it will be the responsibility of the architect/developer to ensure that the relevant Building Control Body (where applicable) is satisfied with the proposals.

# Contents

|  | *Page* |
|---|---|
| Dedication | iv |
| Foreword | v |
| Preface | vi |
| Background | vii |
| Acknowledgements | viii |
| Introduction | 1 |
| Building Regulations and planning policy | 4 |
| Format of the guide | 6 |

## Chapters

| 1 | Strategic site development | 7 |
| 2 | External approach routes and parking | 9 |
| 3 | Communal entrances, doors and circulation | 19 |
| 4 | Individual dwelling entrances and other external doors | 29 |
| 5 | Dwelling circulation areas and storage | 37 |
| 6 | Moving between levels within the dwelling | 47 |
| 7 | Using living spaces | 51 |
| 8 | Using the kitchen | 57 |
| 9 | Bedrooms | 71 |
| 10 | Bathrooms | 79 |
| 11 | Operating internal doors and windows | 99 |
| 12 | Services and controls | 103 |
| 13 | Using outdoor spaces | 109 |
| 14 | Designing wheelchair adaptable dwellings | 113 |

## Appendices

| 1 | Furniture schedule | 125 |
| 2 | Glossary of terms | 127 |
| 3 | Legislation and technical standards | 129 |
| 4 | Organisations and sources of information | 129 |
| 5 | Cross-referencing | 130 |

**Index** 133

Wheelchair Housing Design Guide

# Dedication

Dedicated to the memory of former Habinteg tenant and Board member, Sir Bert Massie (31st March 1949 – 15th October 2017).

We're profoundly grateful for the impact of his hard work promoting accessible homes and inclusive communities, not only on Habinteg, but on the lives of disabled people far and wide.

# Foreword

As Vice Chair of Habinteg and a wheelchair user myself I have seen first-hand the difference that well-designed wheelchair accessible housing can make. Time and again I've heard about the transformative effect that the right home has had on family life, work and training prospects, health and wellbeing.

That's why I'm delighted that Habinteg have collaborated with the Royal College of Occupational Therapists Specialist Section – Housing and the RIBA to publish this important guide. As well as setting out the features of wheelchair accessible housing that are now governed through Building Regulations, it provides best practice recommendations on how these requirements can be delivered and enhanced to produce really liveable homes for a range of wheelchair users.

Crucially, this volume also provides the rationale for each aspect of accessible design – encouraging the reader not simply to implement recommended features but to understand why they are recommended and to think about the likely occupants of the home from the outset. For me this adds huge value, providing the 'story' that will help change perceptions of accessible design and enable a far larger number of practitioners to develop their understanding and intuition.

There is a growing need for wheelchair accessible housing and high-quality design is important across the spectrum, from social rented housing right through to privately owned or even self-build properties. The benefits to the individuals and families are invaluable, whatever the tenure.

I sincerely hope that this guide enriches the skill and ambitions of all those who pick it up.

**Andrew Gibson**
Vice Chair, Habinteg

# Preface

The second edition of the *Wheelchair Housing Design Guide* has been an essential text for occupational therapists working in housing since its publication in 2006, so when, in 2014, the Royal College of Occupational Therapists Specialist Section – Housing (RCOTSS-Housing) was looking for worthwhile projects in which to invest funding, it was no surprise that involvement in the production of a third edition was a popular choice with members. Since 2006 there have been many changes for disabled people, some more positive than others, but one thing that is certain is that there are still too many people living restricted lives in unsuitable and inappropriate housing, and we are happy to make a contribution towards changing that.

RCOTSS-Housing have a long association with Habinteg and we share their philosophy that high-quality accessible housing changes lives for the better. RCOTSS-Housing is delighted to have been able to provide their support and expertise to co-produce this very informative and detailed publication, which will be an invaluable reference point for specialist section members and other professionals working to design and build homes which will enable wheelchair users to live life to the full.

**Gill Owen-John**
Chair – Royal College of Occupational Therapists Specialist Section – Housing
2012–2016

# Background

This third edition of the *Wheelchair Housing Design Guide* is a joint initiative by Habinteg Housing Association with the Royal College of Occupational Therapists Specialist Section – Housing.

Previous editions of this guide have been based on a combination of research undertaken by a cross-section of wheelchair users and the work of housing association practitioners, architects and other professionals. Previous editions therefore provided a strong foundation for the development of *Approved Document M, Volume 1: Dwellings, M4(3) Category 3: Wheelchair user dwellings*, which was published in 2015.

This third edition is therefore based on the M4(3) requirements and aims to guide readers through the regulations in practical terms and offer additional design guidance which professionals should consider when designing homes for wheelchair users.

The authors of this book have knowledge from their backgrounds in architecture, building control and occupational therapy and a wealth of practical experience in the access and disability field. Habinteg warmly welcomed the collaboration between two leading occupational therapists commissioned by the Royal College of Occupational Therapists and two professionals from the Centre for Accessible Environments to author this new edition.

## About Habinteg
Habinteg is a housing association with a unique vision. Founded in 1970, the organisation wants communities to include disabled people, offering places to live that meet their needs and provide the highest levels of independence, choice and control over their daily lives. Habinteg champions inclusion by providing accessible homes and neighbourhoods that welcome and include everyone and by using its expert knowledge to inspire and influence decision makers.

## About the Centre for Accessible Environments
The Centre for Accessible Environments (CAE) pioneered the provision of access guidance for building designers nearly 50 years ago and remains a leading authority on inclusive design. Today, CAE is a technical division within Habinteg, where it continues to provide consultancy, training and publications on designing to meet the needs of disabled and older people wherever they choose to live, work or visit.

## About the Royal College of Occupational Therapists Specialist Section – Housing
The Royal College of Occupational Therapists Specialist Section – Housing is a branch of the Royal College of Occupational Therapists (RCOT). The membership comprises occupational therapists from a variety of backgrounds including educators, researchers, occupational therapy practitioners and students who are interested in promoting good practice in housing, inclusive design and accessible home environments. It considers housing and the built environment as crucial to health and wellbeing. As a Specialist Section it works with RCOT and key strategic partners to influence housing policy and raise design standards.

# Acknowledgements

We are grateful to the following people and organisations who reviewed the content and gave valuable advice during the preparation of this third edition.

Building Control Alliance
Centre for Accessible Environments Associates: Chris Harrowell, Ron Koorm, Helen Allen
RCOT Practice Publications Group
Justin Bannister, PRP Architects LLP
Julie Fleck OBE, Strategic Access Advisor
Jean Hewitt, Centre for Accessible Environments
Philippa Jackson, London Borough of Camden
Ajay Kambo, IDP
Christina McGill, Habinteg
Paraig O'Brien, Department of Health/Northern Ireland Housing Executive
Bernice Ramchandani, The Hyde Group
Trevor Rogers, The Royal Institution of Chartered Surveyors
Dr Rachel Russell, Occupational Therapy Lecturer, University of Salford
Luke Turner, Principal Architect – AD M and AD K, Ministry of Housing, Communities and Local Government (MHCLG)

# Introduction

Welcome to the third edition of the *Wheelchair Housing Design Guide.*

In the two decades since the first edition was published much has changed in terms of standards of accessibility and inclusive design in new housing. This revised edition of the guide provides expert advice on how to design and incorporate good practice into wheelchair accessible homes while meeting current Building Regulations. The guide has been updated in collaboration with the Royal College of Occupational Therapists Specialist Section – Housing to ensure that it meets the requirements of the health professionals who will also be using this valuable resource.

Since Habinteg was established in 1970, we have been synonymous with the principles of accessible housing, independent living and inclusive communities that everyone can enjoy. Our philosophy is based on the social model of disability, so we know that wheelchair accessible homes can help change lives. Previous editions of our *Wheelchair Housing Design Guide* have been widely adopted and have proved influential in ensuring that essential technical detail is clearly understood. This really matters because evidence shows that inclusive housing can help with finding a job, taking up educational opportunities, playing a role in the community and enjoying a full social life – things that many people take for granted.

Demand for wheelchair accessible homes is high.

As the late Sir Bert Massie, former Chair of the Disability Rights Commission, said in his foreword to the last edition:

> *All of us have our own housing stories: first homes, renting, buying, sharing with others and moving to other areas; but for wheelchair users options are often very limited and the stories of the difficulties of ever finding anything suitable distressingly familiar.*

This is as true in 2018 as it was in 2006 and that is why, despite some progress, this design guide is as relevant and important as ever.

## Aim

The purpose of this design guide is to provide guidance and good practice examples of how to design homes which are accessible and useable by wheelchair users, in order to maximise independence.

We hope this guide will be a valuable tool for a range of professionals in the sector, providing insight into the ever-changing field of access and inclusion. We hope that it helps to improve the standard of wheelchair housing and remains a relevant resource for best practice in wheelchair housing and accessible design for years to come.

We believe that no wheelchair user should face barriers to independence as a result of inaccessible design, and are pleased to play a part in stimulating a commitment to well-designed wheelchair accessible homes.

# Introduction

## Scope of the guide

This guide is appropriate for anyone designing new homes for wheelchair users, including designers and developers seeking to meet the requirements of the Building Regulations for England and *Approved Document M, Volume 1: Dwellings, M4(3) Category 3: Wheelchair user dwellings*. It may also be voluntarily used in areas of the country where the optional requirement has not been adopted, or indeed outside England, and is relevant for anyone aiming to build to a wheelchair accessible standard or designing for a specific individual.

Some parts of this guide will also be helpful when considering modifications, adaptations and extensions to an existing home for a wheelchair user.

## Basis for design considerations

This guide aims to outline design considerations for a range of wheelchair users; however, it is acknowledged that there is no typical wheelchair user and wheelchairs vary in size and type. People who need to use a wheelchair in their own homes represent a cross-section of the population, and include children, parents, older people, students or working individuals. Wheelchair users may live alone or with others and some households may have more than one wheelchair user.

Wheelchair users may also have a range of physical impairments and could live independently or require varying degrees of assistance within their home. This guide describes possible difficulties with reach, dexterity and upper limb control that some wheelchair users may experience and makes reference to a range of potential transfer methods.

Although the design considerations try to take these variances into account, it is not possible to comprehensively cover access requirements specific to every wheelchair user.

**Other impairments**
Additional design considerations, such as those to take account of other impairments, are outside the scope of this guide. Some wheelchair users, or people living with wheelchair users, may have access requirements other than physical access. These could include vision or hearing impairments, or perhaps a neurodiverse or cognitive impairment, such as dementia. It is not possible to extend this guide to take account of all potential needs, but we hope that readers will understand the significant value of considering other impairments during the design process. Design guidance for people with vision impairments, for example, may include particular attention to visual contrast between key finishes, while designing for hearing loss will pay increased attention to acoustics and related technology.

**Bespoke design**
Spatial requirements, furniture sizes and access zones are based on average wheelchair sizes and do not take into account larger powered or tilt-in-space wheelchairs or specific considerations for bariatric wheelchair users. Where bespoke wheelchair accessible housing is being designed for a specific individual or the end user is known at the design stage, any additional requirements (over and above the minimum requirements) should be accounted for within the design to avoid significant alterations being required at a later date.

# Introduction

Identifying any specialist equipment that may be required by an individual should be done as early as possible in the planning stages to ensure that equipment can be accommodated and used effectively. Examples of this may include additional space in the bathroom for a height adjustable changing table or selection of a bath suitable for use with a bath lift. Detailed consideration for this degree of bespoke design is not included in the scope of this guide.

**Adaptations to existing homes**
The technical provisions within this guide apply to new build dwellings where the spatial requirements can be factored in from the outset. The design considerations and general principles, however, remain relevant for anyone designing adaptations for a wheelchair user within an existing home. The full spatial and circulatory requirements may not be achievable, but the reasoning behind these requirements can be used to prioritise the use of available space.

**Technology**
The rapid changes in technology and the impact it can have within our homes and on our lives can be of particular benefit to many disabled people. Assistive technology and environmental controls can enable many disabled people to perform daily tasks and make choices independently. Simple household features, such as lighting, heating and curtains, can be controlled remotely, improving independence.

While technology can assist with overcoming many barriers, it is important to remember that this should not be relied on for all functions and may not be suitable for all users.

# Building Regulations and planning policy

## Building Regulations

**Approved Document M, Access to and use of buildings (AD M)**
**Volume 1: Dwellings**

Accessible housing standards were included in Part M of the Building Regulations in October 2015, with supporting guidance in Approved Document M, Volume 1, which applies to all new build dwellings.

Approved Document M, Volume 1: Dwellings has three categories and their application will be dependent on local planning policy and planning conditions:

- M4(1) Category 1: Visitable dwellings is a mandatory baseline requirement that will apply to all new dwellings where a higher optional category is not required.

- M4(2) Category 2: Accessible and adaptable dwellings, is a higher optional standard which may form part of the planning permission conditions. These dwellings provide reasonable access for a wide range of occupants but are not necessarily suitable for full-time wheelchair users.

- M4(3) Category 3: Wheelchair user dwellings is the highest optional standard which may be conditioned as part of the planning permission. This category is divided into two standards:
    - M4(3)(2)(a) – Wheelchair adaptable dwellings, and
    - M4(3)(2)(b) – Wheelchair accessible dwellings.

Planning permission will specify which dwellings need to be designed to be wheelchair accessible or wheelchair adaptable (wheelchair adaptable being the default).

Wheelchair accessible dwellings are designed and constructed to be suitable for a wheelchair user to live in on completion, while wheelchair adaptable dwellings may require some simple adaptations to be suitable for occupation by a wheelchair user.

This guide considers only M4(3) Category 3: Wheelchair user dwellings and the guidance contained in the Approved Document 2015 edition, incorporating 2016 amendments. It expands on, and offers interpretation of the minimum standards in Approved Document M, Volume 1. It includes explanation of how design features aim to meet the needs of wheelchair users in and around their homes, with additional good practice design guidance.

This design guide is appropriate for those designing for all tenures and either M4(3)(2)(a) Wheelchair adaptable dwellings or M4(3)(2)(b) – Wheelchair accessible dwellings. Chapters 1 to 13 are relevant to both wheelchair adaptable dwellings and wheelchair accessible dwellings, unless noted otherwise. Further details on the provisions for wheelchair adaptable dwellings can be found in Chapter 14: Designing wheelchair adaptable dwellings.

**Fully furnished plans**

When making submissions for Building Regulations approval, it is necessary to demonstrate that a dwelling is capable of meeting the functional and spatial provisions of M4(3) Category 3. Furnished plan layouts should be provided to Building Control showing the necessary access zones, other provisions and the prescribed furniture items. Such plans should be shown at a scale of at least 1:100.

# Building Regulations and planning policy

Although these plans are only required for Building Regulations approval, we strongly recommend that they are generated at planning stage in order to ensure that a compliant and useable layout can be achieved.

For ease of reference, the list of furniture and minimum sizes required by M4(3) Category 3 is included in Appendix 1 of this design guide.

## Planning policy

Planning policy is set both nationally, through the National Planning Policy Framework and the National Planning Policy Guidance, and locally, through local plans and planning guidance. Many local plans have adopted the optional requirement M4(3) Category 3: Wheelchair user dwellings for a proportion of all new homes and will require this through the planning process. The planning permission should specify where dwellings need to be designed to meet M4(3)(2)(a) – Wheelchair adaptable dwellings or M4(3)(2)(b) – Wheelchair accessible dwellings.

# Format of the guide

**Principle**
Each chapter is introduced with an overarching principle outlining the expected provision for access, inclusion and usability.

**Design considerations**
The first section of each chapter looks at design considerations, providing explanations, reasoning and aspects that should be taken into account when designing a wheelchair user's home.

**Technical provisions**
The second section outlines the minimum provisions of Approved Document M, Volume 1, M4(3) Category 3 and includes measurements and technical illustrations.

The technical provisions in Chapters 1 to 13 are relevant to both M4(3)(2)(a) – Wheelchair adaptable dwellings and M4(3)(2)(b) – Wheelchair accessible dwellings, unless noted otherwise. Chapter 14: Designing wheelchair adaptable dwellings sets out the areas of difference.

References to M4(3) are made where relevant throughout the guide for ease of cross-referencing (numbers shown indicate the relevant paragraph of the Approved Document; abbreviations: Diag. = diagram, Perf = M4(3) performance criterion, Appx = appendix).

All dimensions on illustrations are minimum unless noted otherwise. Drawings are not to scale.

**Good practice recommendations**
Additional good practice recommendations are given in each chapter. These are not necessary to meet Approved Document M, Volume 1, M4(3) Category 3; however, they should be incorporated into designs where practicable.

# Strategic site development 1

## Principle

Landscaping and levels across a development should be considered at an early stage to ensure a high degree of accessibility for all. Wheelchair user dwellings should be carefully located so that their residents have easy and convenient access, not only to their own dwelling but to the wider site, neighbours, local amenities and public transport facilities.

## 1.1 Design considerations

Clients, developers, architects, designers and planners should carefully consider the design of the site layout at an early stage to ensure that a holistic and inclusive approach to access is incorporated in the initial design.

The location of wheelchair user dwellings should be considered within the overall site layout so that, where practicable, they are evenly distributed (both across the development and within individual blocks of flats), and positioned on the most accessible parts of the site. Many wheelchair users will be regular drivers, car passengers or use some form of accessible transport. It is therefore essential that the external layout of the development and the relationship between dwellings, local amenities, parking and drop-off facilities are carefully planned from the outset. Where practicable, wheelchair user dwellings should be in close proximity to transport links and other facilities beyond the site curtilage.

The design of communal space, such as gardens, landscaped areas, play areas or other facilities (including scooter, cycle or bin stores) should enable access to, and use by wheelchair users.

Careful layout and detailing is often required to resolve differences in level across a site and to ensure that suitable gradients are provided to footpaths. The location of appropriately designed crossing points and dropped kerbs will also be essential to enable wheelchair access around the development and to local amenities.

Slopes and gradients can inhibit ease of movement and should therefore, where practicable, be eliminated or kept short and to shallow gradients. Crossfalls, or falls across the direction of movement, also present difficulties. Such falls should be eliminated, or at least kept to a minimum gradient. Similarly, falls in two directions should be eliminated where possible.

Uneven and rough textured surfaces or loose materials, such as gravel, are difficult for many people to use and can be a particular hazard for wheelchair users. The wheels of a wheelchair can sink into loose gravel, making it difficult to move. Selection and detailing of paving and finishes should ensure that surfaces are firm, even and smooth enough to be wheeled over (unless tactile paving is intentionally used, e.g. tactile warning surfaces).

Wheelchair Housing Design Guide

# 1 Strategic site development

For safety and security, developments should be designed to minimise secluded areas with parking, entrances and paths in clear public view. Lighting should be carefully placed to evenly illuminate outdoor areas and entrances without causing light pollution within the home.

Single-storey wheelchair accessible homes are generally the most easily designed, suitable for wheelchair use and potentially the most space and cost effective. Other possibilities, such as two or three storey dwellings, are feasible but will require the installation of a lift inside the dwelling to enable a wheelchair user to access all floors. This has an impact on the initial installation costs, ongoing maintenance and a significant impact on the space required. Some wheelchair users, however, may still prefer dwellings over more than one storey and providing a choice of dwelling types within the overall development should therefore be considered.

While a ground floor dwelling is convenient for access and egress, not all wheelchair users will want to live (or feel secure living) at ground floor level. In blocks of flats, wheelchair user dwellings should be sited to offer choice on different floor levels. Where flats are located on upper levels, they should be close to communal lifts, with consideration given to emergency egress.

Within a development, wheelchair user dwellings should be a variety of sizes, offering a choice of bedroom and bedspace configurations.

It is also important to remember that wheelchair users may wish to visit their friends and neighbours throughout the development. Ideally, wheelchair access should be available across the whole site, and all homes within the development should be built to a standard that enables this.

# External approach routes and parking

# 2

## Principle

External approach routes from any parking area, drop-off point or the site boundary to communal and private entrances, communal facilities and outdoor spaces, should be accessible and useable by wheelchair users.

## 2.1 Design considerations

Consideration should be given to a wheelchair user's journey from the site boundary to the entrance of their private dwelling, ensuring that all aspects of the route are step-free and wheelchair accessible.

The approach route to a dwelling's entrance will vary between dwelling types and different sites. For example, the approach route to a terraced house may include a driveway, gate or path, while the approach route to a flat may involve one or more communal paths, gates, doors, lobbies, lifts and corridors to reach the entrance. The journey between any drop-off points or parking and the dwelling entrance also forms part of the approach route so travel distances between wheelchair accessible parking and any relevant entrances should be minimised.

Wheelchair accessible dwellings should be located on the most accessible areas of the site to ensure a level or gently sloping approach can be accommodated.

Ramps may be necessary to overcome a difference in level; however, they can be difficult for a wheelchair user to negotiate. Crossfalls, or falls across the direction of movement, also present difficulties. Where ramps are necessary to accommodate a change in level, paths of a suitable gradient and level landings should be provided. The minimum feasible gradient should be used, with crossfalls eliminated where possible. The space required for a ramp and landings is often underestimated and should be considered at early planning stages.

### Figure 2.1 Crossfalls

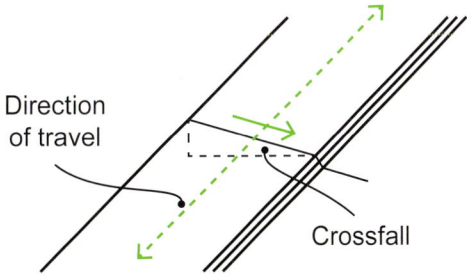

Wheelchair Housing Design Guide

# 2 External approach routes and parking

If there is a change of height between the approach route and the adjacent ground level then appropriate upstands to the path edges should be provided to prevent the wheelchair user veering off the path.

Some wheelchair users are sensitive to rough surfaces so the selection of paving should ensure that approach routes are slip resistant and smooth enough to be wheeled over. Loose materials, such as gravel or sand, should be avoided as they are difficult to push a wheelchair over and can cause skidding. The route should be of a suitable width to allow a wheelchair user easy movement. Where needed, there should be sufficient space to turn or pass other people without leaving the path.

The approach route should be barrier free with gates or doorways kept to a minimum. Where provided, these should be suitably designed to ensure they are easy to use. Doors that open over a landing at the top of stairs or a ramp can be particularly hazardous as a wheelchair user is at risk of falling down the stairs or ramp while negotiating the door swing. Any door or gate swing should therefore be clear of all level landings to ramps and stairs.

Doors and gates should provide clear space, or nibs, to the leading and following edge of the door. This enables a wheelchair user to reach the handle by providing space alongside to accommodate the protrusion of a person's feet and wheelchair footplates. Doors and gates should require minimal force to open. Gates should be aligned with other external doors to assist with head-on approach and preferably open in the same direction.

Windows should be positioned or specified to ensure that they do not cause an obstruction. Outward-opening windows over paths and gardens may present a hazard, particularly to wheelchair users and children.

### Figure 2.2 Window positioning

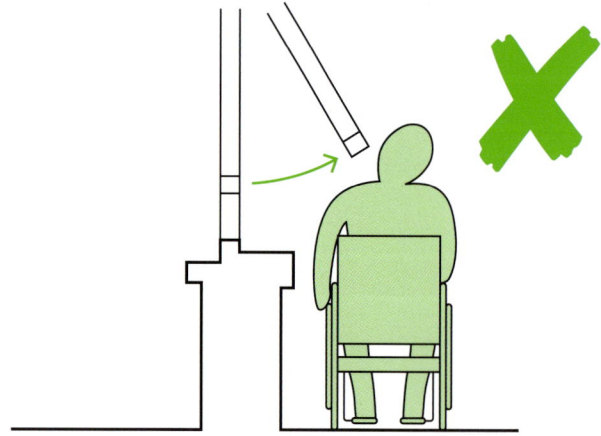

# External approach routes and parking

It is important that a wheelchair user is able to approach and use communal storage areas, such as refuse, cycle or mobility scooter stores. It should be possible to independently deposit refuse and materials for recycling in containers for collection. Thought should be given to the distances a wheelchair user would have to travel to reach these facilities and distances should be minimised were possible. Any facilities located adjacent to the approach route should be carefully designed to ensure they do not create a hazard or obstruction.

All other external spaces, such as communal gardens, terraces or play areas, should be accessible for a wheelchair user to enjoy. Seating and other street furniture around the development should be located so that they do not impinge on footpaths.

All approach routes to facilities and entrances should be well lit. The positioning of lighting columns and bollards should ensure consistent lighting levels, while avoiding glare.

**Car parking**

Transferring into and out of a vehicle can be time consuming and it is important that wheelchair users are able to stay dry. Consideration should be given to effective weather protection between the dwelling entrance and parking space, with a style that is consistent with the design of the overall development.

Wheelchair users may use a variety of methods to transfer into their car and transport their wheelchair. Car parking should therefore be appropriately sized to enable a range of transfers.

Space alongside the car allows a wheelchair user to transfer laterally from their wheelchair directly into the car. Space may also be required to the rear of the car to access the boot for wheelchair storage or to accommodate a boot hoist. The wheelchair may also be stored in a rooftop box, requiring adequate height clearance in enclosed or covered parking spaces.

There are a variety of wheelchair accessible vehicles, which enable a wheelchair user to travel as a passenger or a driver without needing to transfer out of their wheelchair. Entry into the vehicle may be from the side or the rear via a ramp or lifting platform, and suitable space is required to accommodate entry and exit. Wheelchair accessible vehicles have a higher roof to accommodate an occupied wheelchair and will therefore require adequate height clearance in covered parking spaces.

### Figure 2.3   Access to parking

Wheelchair Housing Design Guide

# 2 External approach routes and parking

To enable transfer into and out of a vehicle, the parking bay and access zone should be kept free of all obstructions, such as columns, bollards and charging points. It is also important that the access zone to parking bays is additional to any pathways or approach routes.

Communal car parking serving multi-storey or low-rise high-density developments should ensure that any wheelchair accessible parking bays are located as close to the relevant principal entrance or lift core as possible. Parking provided in basement or undercroft parking, or any parking accessed via a vehicle lift or ramp, should provide adequate height clearance.

Private car parking spaces should be located as close to the principal private entrance as possible. Where carports are provided, care should be taken to limit any loss of daylight affecting any adjacent home and the carport should ideally not be positioned in front of windows.

Private garages, owing to their confined nature, are less accessible than carports. Where provided, ensure that the door provides adequate height clearance and can be operated remotely.

Wheelchair users may also sometimes use a mobility scooter to travel longer distances. Consideration should therefore be given to providing suitable scooter storage, and a charging point under cover, within easy access of the dwelling entrance or a relevant lift.

**Drop-off points**

Drop-off points should be located as close to the relevant dwelling entrances as possible. They should be designed to enable access to either side and to the rear of the vehicle to accommodate a range of wheelchair accessible transport. Dropped kerbs and an accessible route should also be provided to access the dwelling.

# Technical provisions

## 2.2 Technical provisions

*When designing to Part M of the Building Regulations, Approved Document M, Volume 1, M4(3) Category 3, the technical provisions detailed in this chapter are relevant to both M4(3)(2)(a) – **Wheelchair adaptable dwellings** and M4(3)(2)(b) – **Wheelchair accessible dwellings**.*

### 2.2.1 Approach route

| | |
|---|---|
| Regardless of which storey the dwelling is located on, an accessible step-free approach route suitable for a wheelchair user should be provided to all private entrances from the parking area, the location that a wheelchair user would get out of a vehicle, or any other defined starting point (typically the pavement immediately outside the curtilage or plot boundary). These points of access may be within or outside the plot of the dwelling or building containing the dwelling. Provisions do not apply beyond the curtilage of the development. | 3.7<br>3.8<br>3.3<br>3.2 |
| For a house, the approach route will often only involve a driveway and a gate or a path. For a dwelling within a larger building (typically a block of flats), the approach route usually involves one or more communal gates, paths, entrances, doors, lobbies, corridors and access decks as well as communal lifts and stairs. | 3.5 |
| Reasonable provision should also be made for a step-free approach route suitable for a wheelchair user to any communal facilities intended to serve the dwelling. Such communal facilities may include refuse stores or other communal storage areas, communal gardens and terraces but does not include plant rooms or other service areas, unless residents need regular access, e.g. to read meters. Approach routes to dedicated storage for mobility scooters (where provided) should also be step-free. | Perf – a<br><br>3.4<br><br>3.7 |
| The approach route should be illuminated by fully diffused lighting activated automatically by a dusk to dawn timer or by detecting motion. | 3.9f |
| Approach routes (whether private or communal) should have a minimum clear width of 1200mm, with a maximum crossfall of 1:40. External parts of the approach route should have a suitable ground surface. These should provide an even, slip resistant surface with no loose laid materials. A level space with a minimum width and depth of 1500mm should be provided at the end of the approach route, and at maximum intervals of 10m to enable passing or turning. | 3.9b<br>3.9e<br>3.9d |
| Localised obstructions, such as columns, planting or street furniture, should be no longer than 2m in length and no more than 150mm deep. Obstructions should not be positioned opposite or close to a doorway or change of direction. | 3.9c |
| The approach route should be safe and convenient for everyone, be at the shallowest gradient that can reasonably be achieved and be step-free. The approach route should be level, gently sloping or ramped. | 3.7<br>3.9a |

### 2.2.2 External (and internal) ramps forming part of an approach route

| | |
|---|---|
| Where a ramped approach is provided, the gradient should be between 1:20 and 1:15, with the maximum length of each flight in line with Table A. | 3.10a<br>3.10b |

Wheelchair Housing Design Guide

# 2 Technical provisions

AD M
Diag 3.1

| Table A | External and internal ramps | | |
|---|---|---|---|
| | Gradient | Rise (mm) | Maximum length of ramp flight (m) |
| | 1:20 | 500 | 10 |
| | 1:19 | 474 | 9 |
| | 1:18 | 444 | 8 |
| | 1:17 | 412 | 7 |
| | 1:16 | 375 | 6 |
| | 1:15 | 333 | 5 |

(steeper ↔ shallower)

Note 1: Gradient × length of flight = rise (e.g. 1/20 × 10 = 500mm).
Note 2: Ramps should be no steeper than 1:15.

Each ramp flight (whether within a private or communal approach route) should have a minimum clear width of 1200mm, with level landings at the top and bottom of each flight. Intermediate level landings are required between individual flights and at any change of direction.

3.10c
3.10d
3.10f
3.10e

Every landing should be level and a minimum 1200mm clear of any door or gate swing. Where a ramp leads directly to a principal private or principal communal entrance, the level landing outside the entrance should be a minimum width and depth of 1500mm and clear of any gate or door swing.

3.10f
3.14a
3.22a

### Figure 2.4  Ramps forming part of an approach route

Diag 3.1
3.10d

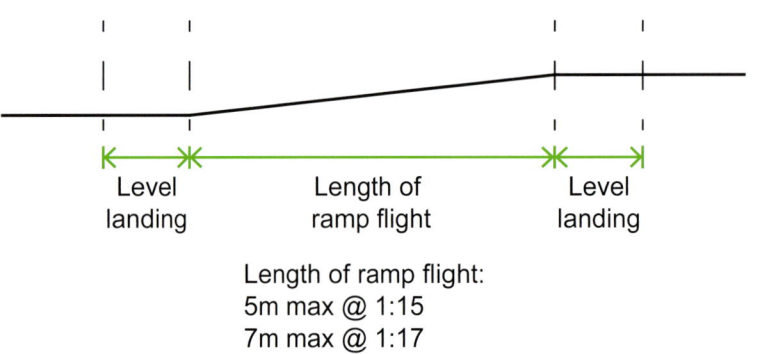

Length of ramp flight:
5m max @ 1:15
7m max @ 1:17
10m max @ 1:20

## 2.2.3 External steps forming part of an approach route

Where there is a communal ramped approach with an overall rise of 300mm or more, an additional stepped route should be provided that enables a wide range of people to use it safely. External steps forming part of an additional approach route should have suitable tread nosings, with uniform risers of between 150mm and 170mm and uniform goings of between 280mm and 425mm. The goings on tapered steps should be measured at a point 270mm from the 'inside' (narrow end) of the step. Single steps should be avoided.

3.8
3.11
3.11b
3.11a

3.11g

# Technical provisions

Every flight of steps should have a minimum clear width of 900mm and individual flights should have a rise of no more than 1800mm between landings. Top, bottom and, where necessary, intermediate landings should be provided and every landing should be a minimum of 900mm long.

Every flight which has three or more risers, should have a suitable handrail on one side of the flight, or on both sides where the flight is wider than 1000mm. The handrail should be easy to grip and be set at a height of 850mm to 1000mm above the pitch line of the flight, extending at least 300mm past the top and bottom nosings.

### 2.2.4 Doors and gates

Every door, gate or gateway along an approach route should provide a minimum clear opening width of 850mm with a minimum 300mm nib to the leading edge (pull side) and a minimum 200mm nib to the following edge (push side). Further requirements, as detailed in 3.2.3, should also be met.

**Figure 2.5 Doors and gates**

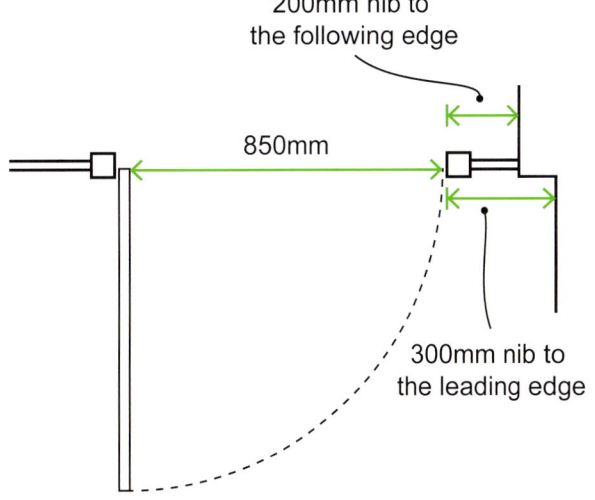

### 2.2.5 Parking bays

Where a dwelling has a parking space, it should enable a wheelchair user to get into and out of a car from both sides and to access the rear of the vehicle. The parking bay should be level. The gradient along the length of the parking space should not exceed 1:60 and the crossfall should not exceed 1:40. Parking bays should also have a suitable ground surface. The surface material should be firm, even, smooth enough to be wheeled over and not be covered in any loose laid materials (e.g. gravel).

> **Additional good practice recommendations**
>
> - Provide a covered parking bay for every wheelchair accessible property. Where this is not located within the curtilage of the dwelling, it should be located as close as practicable to the principal entrance or nearest lift to reach this entrance.
>
> - Provide weather protection from the wheelchair accessible parking to the dwelling entrance. Cover should be in keeping with the overall design of the development or dwelling.

Wheelchair Housing Design Guide

# 2 Technical provisions

| | AD M |
|---|---|

### 2.2.6 Private parking bays

Where a parking space is provided within the private curtilage of a dwelling (including a carport or garage) it should be 2400mm wide × 4800mm long and have an additional minimum 1200mm clear access zone to one side and to the rear. There should be minimum clear headroom of 2200mm.

3.12a
3.12d

### Figure 2.6   Private garage

3.12a
3.12d

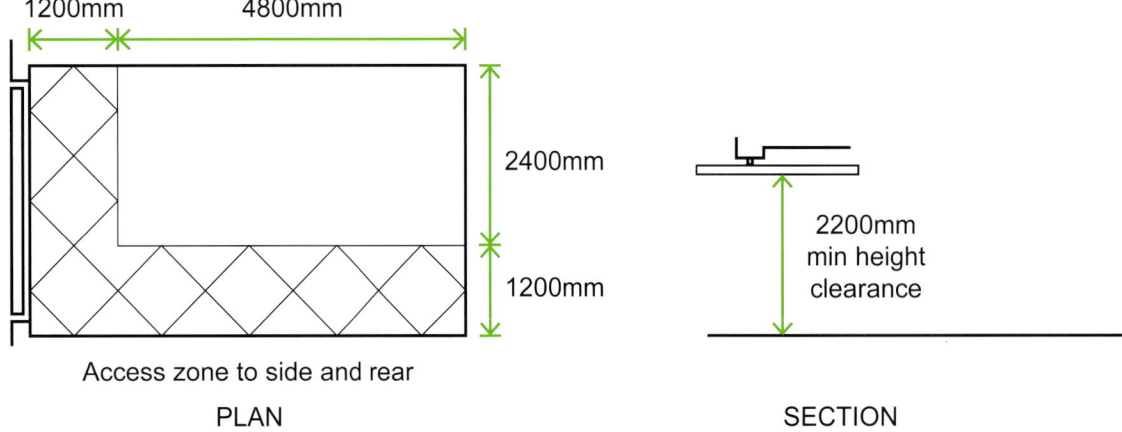

Access zone to side and rear
PLAN                                       SECTION

### 2.2.7 Communal parking bays

Where a parking space is provided within a communal parking area it should be 2400mm wide × 4800mm long and have an additional minimum 1200mm clear access zone to both sides. The side access zones can be shared between two communal parking bays. A minimum clear headroom of 2200mm is required.

3.12b
3.12 Note
3.12d

### Figure 2.7   Communal parking bays

3.12b
3.12 Note

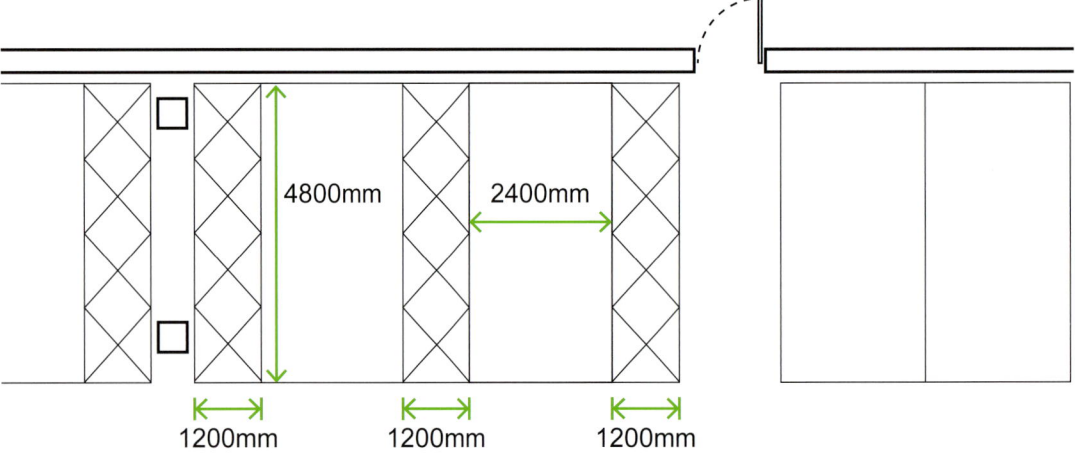

Access zones clear of obstructions

16                                                                                   Wheelchair Housing Design Guide

# Technical provisions

### 2.2.8 Drop-off points

Where a drop-off or setting down point is provided for the dwelling, it should be level with a gradient not exceeding 1:60 and a crossfall not exceeding 1:40. The drop-off point should have a suitable ground surface. The surface material should be firm, even, smooth enough to be wheeled over and not be covered in any loose laid materials (e.g. gravel).

The drop-off or setting down point should be located close to the principal communal entrance of the core of the building containing the dwelling. Where a dropped kerb is provided, it should be a minimum of 1000mm wide, be reasonably flush with the adjoining ground and have a maximum gradient of 1:15.

**AD M**

3.13
3.13b
3.13c

3.13a
3.13d

# Communal entrances, doors and circulation    3

## Principle

Communal entrances, circulation areas and facilities should be well designed and provide adequate manoeuvring space for wheelchair users, their personal assistants, family and visitors to access and negotiate the building and its facilities.

## 3.1 Design considerations

The design of communal entrances, doors, corridors and communal facilities should be carefully considered to enable ease of access. The approach route through communal areas to all wheelchair user dwellings and associated facilities should be wheelchair accessible. Travel distances to wheelchair accessible dwellings from any principal entrance and/or lifts should be minimised.

Communal entrance doors should have suitably sized level landings, clear of door swings, to ensure that wheelchair users can stop safely at the door to unlock, open and manoeuvre through it. A canopy or other form of weather protection should be provided.

Doors should be well lit externally with lighting that illuminates all the key elements: lock, handle, entry phone, bell, door number and letter box. Lighting that is automatically activated by a dusk to dawn timer or by detecting motion should be provided.

A heavy entrance door, or one fitted with a door closer, can be difficult for a wheelchair user to open and to hold open while moving through the doorway. Where the recommended opening force cannot be met, a power assisted door should be provided.

Door thresholds and door frames should provide an accessible threshold. Even minimal upstands create a barrier to movement and are difficult to propel a wheelchair over, so these should be designed out.

A wheelchair user may approach a closed door at an angle to enable them to reach the door handle. Additional space, or clear door nibs, are required to the leading edge (pull side) and following edge (push side) of the door to accommodate feet and footplate protrusion. Where the door opens in the direction of travel, the manoeuvre is simpler than if it opens towards the wheelchair user. The additional width created by these nibs should be maintained for an appropriate distance to enable a wheelchair user to reverse or move through the door, clear of the door swing. The wider the nib is, the more space there is available for these manoeuvres and the easier the door is to negotiate.

# 3 Communal entrances, doors and circulation

Figure 3.1 Door nibs

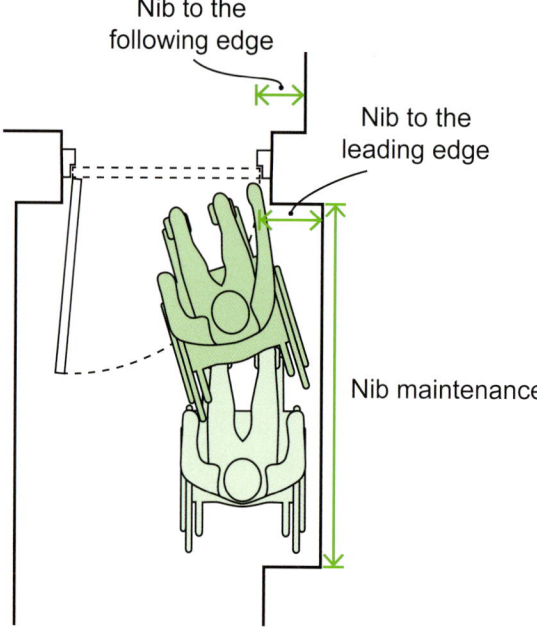

Doors should be located reasonably centrally within the thickness of the wall to minimise the depth of the reveal to both sides. Where the reveal is deep, the structural opening should be enlarged to accommodate the required nib next to the door handle, enabling a wheelchair user to approach and operate the door.

Figure 3.2 Door nibs with a deep reveal

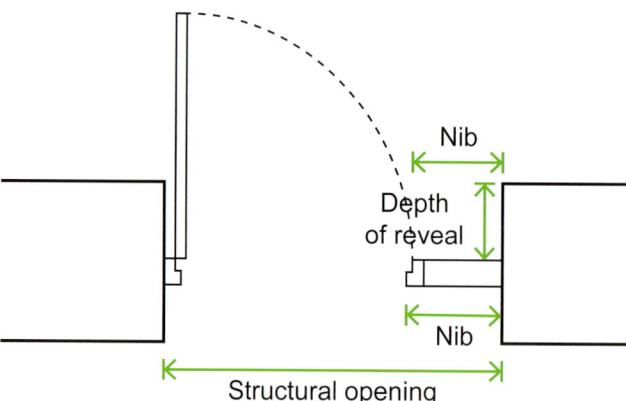

Communal doors, while necessary for fire or security reasons, can be a barrier to wheelchair users, particularly when heavy or fitted with poorly maintained door closers. The number of communal doors on any route or to reach any communal facilities, such as a bin, cycle or scooter storage area, should therefore be reduced to the minimum necessary.

The use of door closers should be kept to a minimum as they can make doors heavy to operate and can result in doors closing too quickly. Where they are necessary, they should include delayed action closing as wheelchair users may take more time to move through the doorway.

# Communal entrances, doors and circulation    3

Where a door does not need to be closed other than to prevent fire spread, consideration should be given to the use of fail-safe hold open devices in line with relevant Regulations. Where doors across corridors are held open, these should be recessed into the wall to minimise localised obstructions and not reduce clear corridor width.

Doors should have an appropriate clear opening width to allow easy access for a wheelchair user. Where there is a double door, the main leaf should provide the minimum clear opening width, as it is difficult for a wheelchair user to open two doors simultaneously.

Where a close proximity control is required to operate a door, consideration should be given to providing a remote key fob as not all users will be able to reach and operate wall-mounted controls.

Vision panels should be provided to doors where privacy is not required. These should be tall and start low enough to provide an adequate sightline through the door.

Where there is an entrance lobby or an additional door beyond the entrance door, there should be sufficient space between the door swings to enable a wheelchair user to pass through one door before negotiating the second.

## Figure 3.3    Entrance lobby

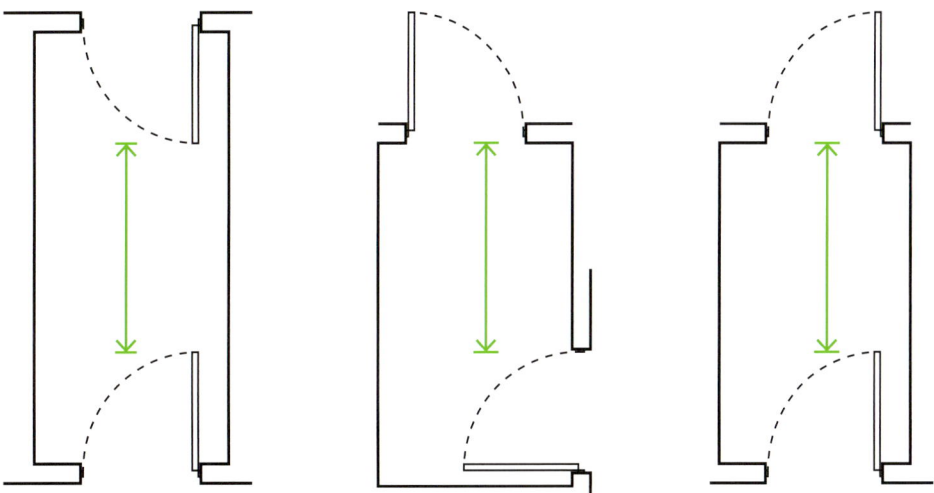

Manoeuvring space between lobby doors and door swings

Suitable flooring or matting, of sufficient length, should be provided at entrance doors to assist wheelchair users in removing dirt and water from their wheels. This should be firm and recessed to ensure a level surface is maintained. Coir matting is unsuitable as it can cause wheels to sink and becomes uneven after time.

The position of post boxes or meter cupboards in the entrance lobby or circulation space should not restrict access or reduce the size of door nibs. Post boxes allocated to wheelchair accessible dwellings should be set at an accessible height.

Wheelchair users require additional space to turn and to pass other people and wait for lifts.

Where corridors need to overcome a change of height, ramps of a suitable gradient should be provided.

Wheelchair Housing Design Guide

# 3 Communal entrances, doors and circulation

**Communal lifts**

Communal lifts should be large enough to accommodate a range of wheelchairs users and their companions. There should preferably be two communal lifts available to access all wheelchair user dwellings so that, in the event of breakdown or servicing, one lift is always available.

Where it is not feasible to provide two lifts, a robust maintenance plan should be in place to ensure lifts are repaired swiftly, enabling residents to leave and access their homes.

# Technical provisions

## 3.2 Technical provisions

*When designing to Part M of the Building Regulations, Approved Document M, Volume 1, M4(3) Category 3, the technical provisions detailed in this chapter are relevant to both M4(3)(2)(a) – **Wheelchair adaptable dwellings** and M4(3)(2)(b) – **Wheelchair accessible dwellings**.*

### 3.2.1 Principal communal entrance door

The principal communal entrance door is the communal entrance to the core of the building containing the dwelling which a visitor not familiar with the dwelling would normally expect to approach (usually the common entrance to a block of flats). The principal communal entrance should enable access by a wheelchair user. — 3.14

A level landing with a minimum width and depth of 1500mm should be provided outside the principal communal entrance, clear of any door swing or obstruction. The landing should be covered for a minimum width and depth of 1200mm. — 3.14a, 3.14b

### Figure 3.4   Principal communal entrance

*Diag 3.2, 3.14a, 3.14b*

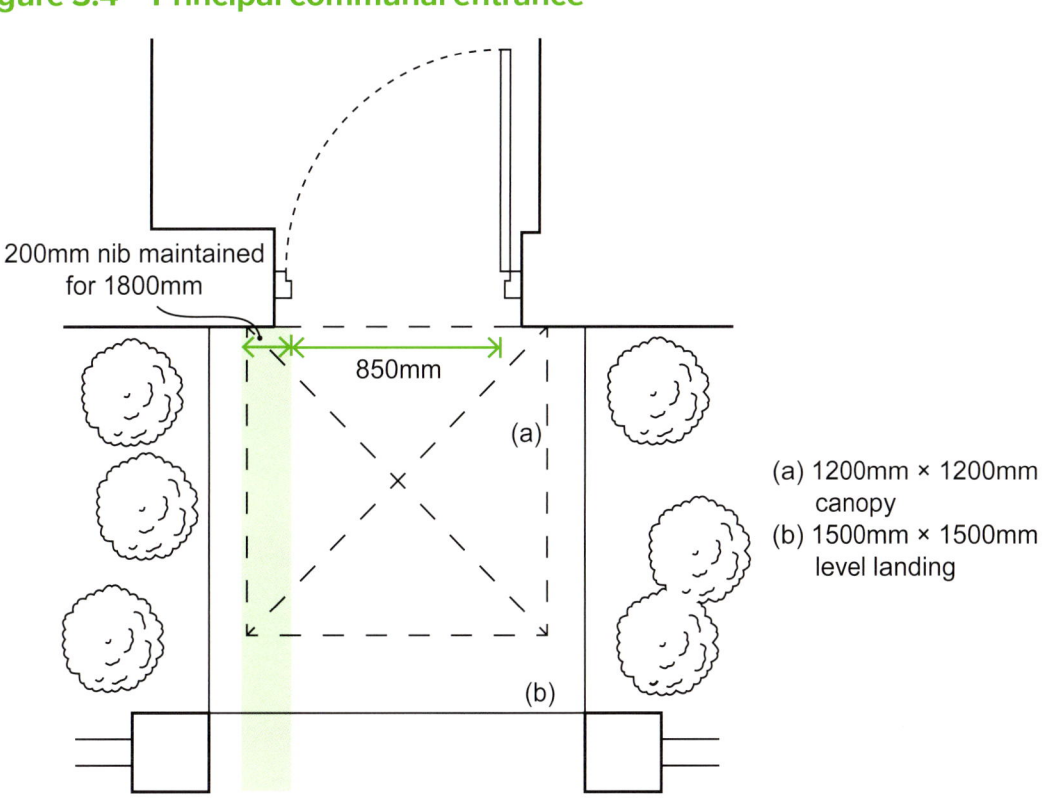

- 200mm nib maintained for 1800mm
- 850mm
- (a) 1200mm × 1200mm canopy
- (b) 1500mm × 1500mm level landing

Lighting should be provided that uses fully diffused luminaires and is activated automatically by a dusk to dawn timer or by detecting motion. — 3.14c

The principal communal door entry controls should be mounted 900mm to 1000mm above finished ground level and a minimum of 300mm away from any projecting corner. A door entry phone with remote door release facility should be provided in the main living space and principal bedroom. Any ironmongery such as handles, locks, latches and catches should be easy to grip and use and fitted 850mm to 1000mm above ground level. — 3.14n, 3.44i, 3.44f

Wheelchair Housing Design Guide

# 3 Technical provisions

There should be sufficient space alongside the door for a wheelchair user to pull up and reach the door handle:

- Provide a minimum 300mm nib to the leading edge (pull side) of the door (or gate), with the extra width created by this nib maintained for a minimum distance of 1800mm beyond it.

- Provide a minimum 200mm nib to the following edge (push side) of the door (or gate), with the extra width created by this nib maintained for a minimum distance of 1800mm beyond it.

The entrance door should be located reasonably centrally within the thickness of the wall while ensuring that the depth of the reveal on the leading face of the door (usually the inside) is a maximum of 200mm.

Provide a minimum clear opening width of 850mm. Where there are double doors (or gates), the main (or leading) leaf should provide the required minimum clear opening width of 850mm.

A clear turning circle of a minimum 1500mm diameter should be provided inside the entrance area, behind the entrance door when closed. Where there is a lobby or porch, the doors should be a minimum of 1500mm apart and have a minimum of 1500mm clear space between door swings.

Circulation space and nibs to doors should be clear of all obstructions, including any post boxes or meter cupboards.

**AD M**

3.14g

3.14h

3.14i

3.14e
3.14f

3.14d
3.14k

3.14d
3.14g
3.14h
3.14k
Diag 3.2
3.14d
3.14e
3.14g
3.14h
3.14i
3.14k

### Figure 3.5 Principal communal entrance

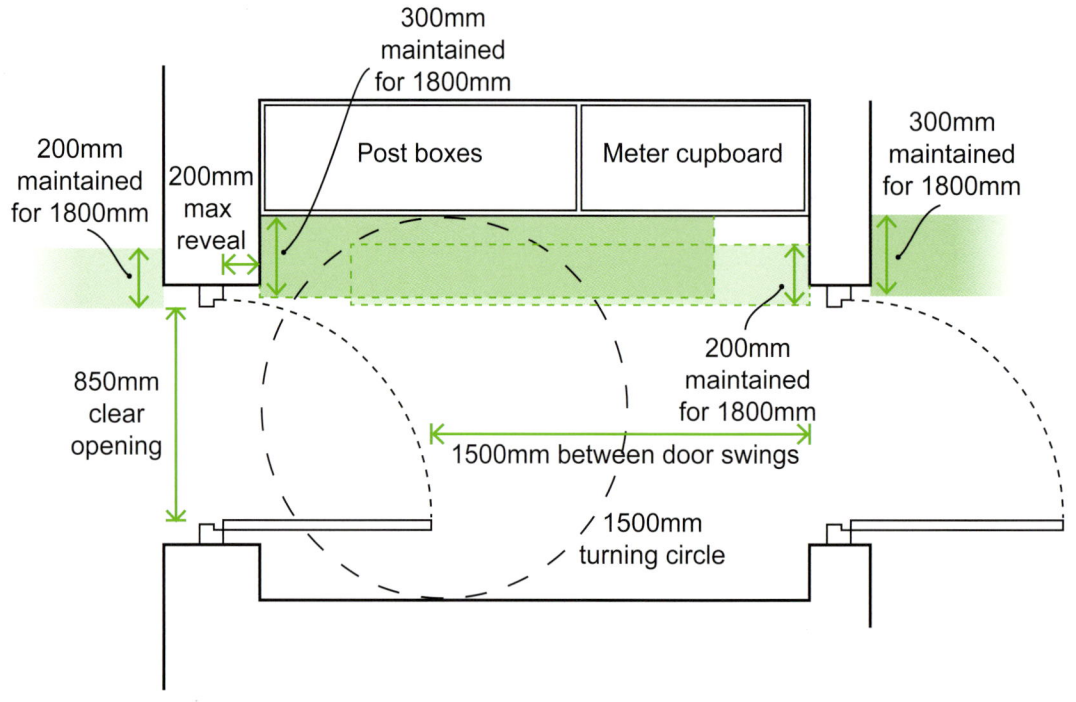

24  Wheelchair Housing Design Guide

# Technical provisions

Doors should have an accessible threshold, i.e. the threshold should be level or, if raised, have a total height of no more than 15mm, a minimum number of upstands and slopes and any upstands higher than 5mm chamfered.

### Figure 3.6  Accessible threshold

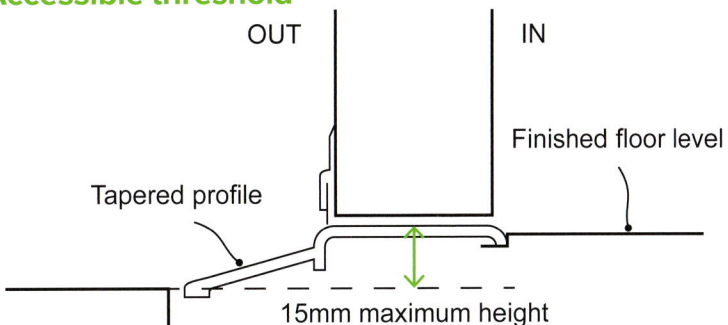

Where the force required to open the door exceeds 30N from 0° to 30° or 22.5N from 30° to 60° of the opening cycle, a power assisted door should be provided.

Suitable ground surface or entrance flooring which does not impede wheelchair movement should be provided at entrances.

> **Additional good practice recommendations**
> 
> - Entrance lighting should clearly illuminate the approach route, entrance door and key elements of the door entry controls.
> - The entrance cover should be a maximum height of 2300mm for effective weather protection, with coverage extending beyond the door on the lock side.
> - Door entry controls should:
>   - have an appropriately positioned camera where video entry is included, that provides a good view of all callers
>   - have call buttons that are easy to identify and operate with reduced dexterity and strength
>   - include visual indicators which indicate response
>   - enable a wheelchair user to open the principal communal entrance and any secure communal doors on the approach route to allow visitors entry without leaving their dwelling.
> - Matting, where provided, should be firm and recessed so the mat is level with the adjacent flooring and creates no upstands or lip. If surface laid, matting should have a rubber backing and chamfered edges to reduce the risk of trips and falls. Deep pile carpets or coir matting should not be used.
> - To assist dirt clearance from the wheels of a wheelchair, provide an entrance mat with a minimum length of 2.1m in the direction of travel.
> - Position post boxes allocated to wheelchair accessible dwellings with their centre line between 700mm and 1000mm above floor level and a minimum of 700mm from an inside corner.

Wheelchair Housing Design Guide

# 3 Technical provisions

| | AD M |
|---|---|

### 3.2.2 Communal corridors

Communal corridors forming part of the approach route to wheelchair user dwellings, and any communal facilities associated with the dwelling, should be wheelchair accessible.

Perf – a
3.4
3.9

These routes should be step-free and be at the shallowest gradient that can reasonably be achieved. Where internal ramps are unavoidable, they should meet the requirements of 2.2.2.

3.7
3.10

Communal corridors should have a minimum clear width of 1200mm, with corridors in front of individual private entrance doors and lifts 1500mm wide, to enable the provision of a clear level landing, a minimum of 1500mm long and 1500mm wide. Where there are long corridors, a level space with a minimum width and depth of 1500mm for turning or passing should be provided at maximum intervals of 10m.

3.9b
3.16a
3.22a
3.9d

Any localised obstructions, such as radiators, meter cupboards, post boxes or held open doors, should be no longer than 2m in length and should not reduce the required corridor width by more than 150mm. Localised obstructions should not occur opposite or close to a doorway or at a change of direction.

3.9c

> **Additional good practice recommendations**
> - Suitable flooring which does not impede movement by wheelchair users should be provided throughout communal approach routes.

### 3.2.3 Communal doors and gates

Every communal door or gate along the approach route to wheelchair user dwellings, and any communal facilities associated with the dwelling, should be wheelchair accessible.

3.15
3.4

There should be sufficient space alongside the door for a wheelchair user to pull up and reach the handle:

- Provide a minimum 300mm nib to the leading edge (pull side) of each door, with the extra width created by this nib maintained for a minimum distance of 1800mm beyond it.

3.15
3.14g

- Provide a minimum 200mm nib to the following edge (push side) of each door, with the extra width created by this nib maintained for a minimum distance of 1800mm beyond it.

3.15
3.14h

Doors should be located reasonably centrally within the thickness of the wall so that the depth of the reveal on the leading face of the door (usually the inside) is no greater than 200mm.

3.15
3.14i

All communal doors should have an accessible threshold, i.e. the threshold should be level or, if raised, should have a total height of no more than 15mm, a minimum number of upstands and slopes and any upstands higher than 5mm chamfered. Doors should provide a minimum clear opening width of 850mm. Where these are double doors, the main (or leading) leaf should provide the required minimum clear opening width of 850mm.

3.15
3.14j
3.14e
3.14f

Doors forming lobbies, porches or across a corridor should be a minimum of 1500mm apart, with a minimum 1500mm clear space between door swings.

3.15
3.14k

# Technical provisions

| | AD M |
|---|---|
| Any door where the opening force required exceeds 30N from 0° to 30° or 22.5N from 30° to 60° of the opening cycle, should have power assisted door controls. | 3.15<br>3.14l |
| Communal door entry controls should be mounted 900mm to 1000mm above floor level and a minimum of 300mm away from any projecting corner. Any ironmongery such as handles, locks, latches and catches, should be easy to grip and use and fitted 850mm to 1000mm above floor level. | 3.15<br>3.14n<br>3.44f |
| Suitable flooring should be provided which does not impede wheelchair manoeuvrability. | 3.15<br>3.14m |

> **Additional good practice recommendations**
> - Vision panels should be provided in doors wider than 450mm, with a visibility zone between 500mm and 1500mm from finished floor level. These should be located towards the leading edge of the door.
> - The specification of ironmongery should ensure that the style reduces the risk of catching clothing, e.g. a 'D' handle.

### 3.2.4 Communal lifts

| | |
|---|---|
| To be suitable for use by a wide range of people, including accompanied wheelchair users, every communal passenger lift leading to a wheelchair user dwelling, should have a minimum lift car size of 1100mm wide and 1400mm deep. | 3.16c |
| A clear landing, a minimum of 1500mm long and 1500mm wide, should be provided directly in front of the lift at every floor level. | 3.16a |
| The lift should be equivalent to, or comply with, the requirements of BS EN 81-70:2003 for a type 2 lift. | 3.16b |
| Lift doors should have a minimum clear opening width of 800mm. The doors also require a five second initial dwell time once fully open, before they begin to close. | 3.16d<br>3.16f |
| Both landing and car controls should be located 900mm to 1200mm above the car floor and a minimum of 400mm (measured horizontally) from the inside of the front wall. | 3.16e |

> **Additional good practice recommendations**
> - Wheelchair user dwellings should preferably be accessed by two communal passenger lifts so that, in the event of breakdown or lift servicing, one lift is always available.
> - Where it is not feasible to provide two lifts, a robust maintenance plan should be in place to ensure lifts are repaired swiftly, enabling wheelchair users to leave and access their homes.

### 3.2.5 Communal staircases

| | |
|---|---|
| The principal communal stair that gives access to a wheelchair user dwelling should be designed as a general access stair to meet the provisions of Part K. | 3.17 |

Wheelchair Housing Design Guide

# Individual dwelling entrances and other external doors 4

## Principle

A wheelchair user should be able to independently operate and negotiate all entrance doors into their property and to private external spaces.

## 4.1 Design considerations

All doors into and out of the dwelling should be well designed with adequate manoeuvring space to enable wheelchair users to operate and negotiate the door.

It can be time consuming and more difficult for wheelchair users to unlock and manoeuvre through an entrance door. An appropriately sized level landing, with weather protection, is therefore essential to ensure that a wheelchair user can stop outside the door, unlock it and enter their home safely and conveniently.

Effective lighting at all entrances that clearly illuminates the lock and handle can assist ease of use and increase the feeling of security. Lighting at the principal entrance should be automatically activated by a dusk to dawn timer or by detecting motion, with the light switch to internal lighting conveniently located by the door. Where there is a secondary approach route from any parking, storage or refuse area, this should also be lit with automatically activated lighting.

A wheelchair user may approach a closed door at an angle to enable them to get close enough to reach the door handle. Additional space, or clear door nibs, are therefore required to the leading edge (pull side) and following edge (push side) of the door to accommodate feet and footplate protrusion. The additional width created by these nibs should be maintained for an appropriate distance to enable a wheelchair user to reverse, open and move through the door. These clear door nibs can also assist in providing enough space for a wheelchair user to move through the door before turning to close it or to move past someone holding the door open for them.

# 4 Individual dwelling entrances and other external doors

Figure 4.1 Door nibs

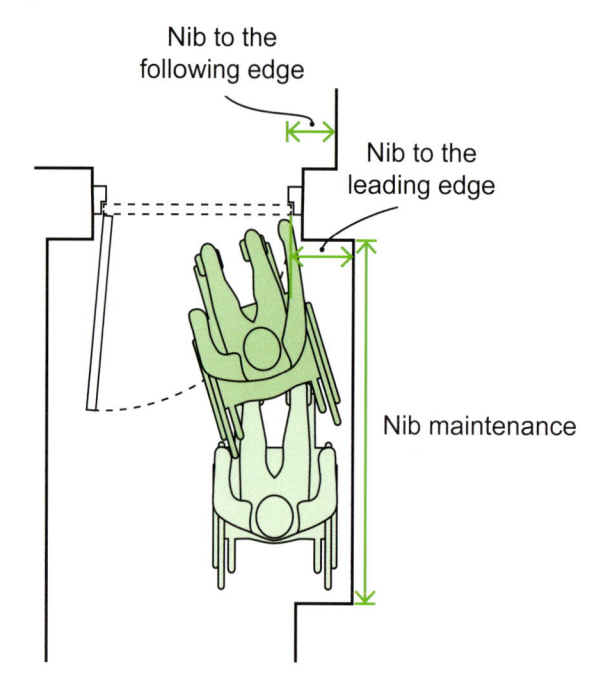

Where sliding doors are used, such as to and from a garden or onto a balcony, adequate space for feet and footplate protrusion is still required to approach the door at an angle and reach the door handle. Both sides of a sliding door should be treated as a leading edge to ensure there is adequate space when approaching the door from either side.

Figure 4.2 Nibs to sliding doors

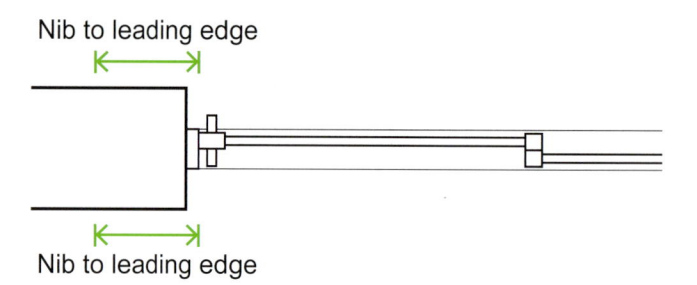

Doors should be located reasonably centrally within the thickness of the wall to minimise the depth of the reveal to both sides. Where the reveal is deep, a wheelchair user may not be able to get close enough to reach the door handle. Where this is the case, the structural opening should be enlarged to accommodate the required nib to the door handle, enabling a wheelchair user to operate the door (refer to Figure 3.2).

Door handles and locks should allow for single handed operation. The handle should not need to be operated (e.g. lifted) while simultaneously locking or unlocking the door.

Door closers can make doors heavy to operate and result in doors closing too quickly. Power assisted opening may sometimes need to be fitted to meet individual requirements.

# Individual dwelling entrances and other external doors | 4

Matting inside the entrance door can assist wheelchair users to remove dirt and water from their wheels. Where provided, matting should be firm and recessed to ensure a level surface is maintained. The durability and ease of cleaning of flooring inside the entrance area should be taken into account. Coir matting or carpeting in this area should be avoided.

Porches or entrance lobbies are only practical if they provide sufficient space for a wheelchair user to clear one door swing before negotiating the other door (refer to Figure 3.3). Where these areas are likely to be used for coats, shoes and other storage, additional space should be factored in to ensure circulation is not compromised.

Passive surveillance at an entrance door can assist in identifying callers. This could include glazing within the door, or to one side, door viewers or overlooking windows, e.g. from bay windows in the living room. Where door viewers are provided they should be positioned at both an upper and lower height.

Figure 4.3    Passive surveillance

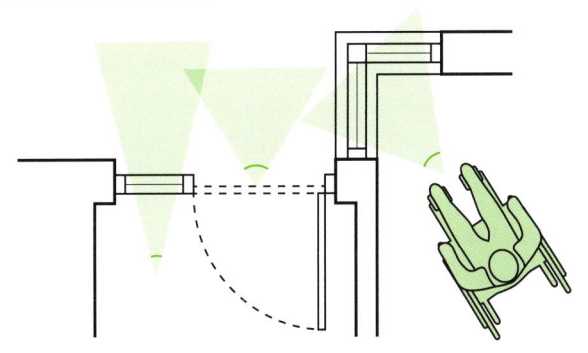

A door entry system, with an appropriately sited camera and display screen, provides a view of callers. Remote door release functions in the living area and principal bedroom assist wheelchair users who may find it more difficult and need more time to get to the door and open it.

It should be possible to approach the door to receive deliveries, retrieve post, open the door to visitors and to turn around. Visitors may also be wheelchair users themselves, so additional space inside the entrance is desirable.

Figure 4.4    Entrance circulation

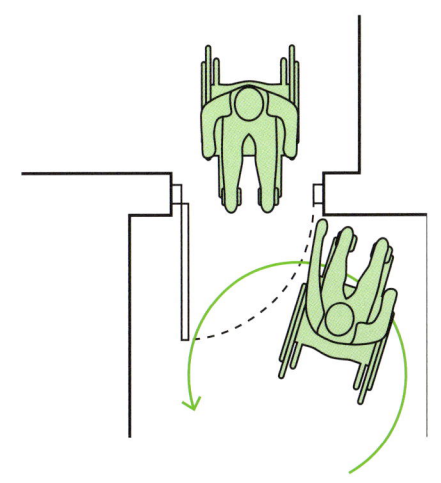

Wheelchair Housing Design Guide

# 4 | Individual dwelling entrances and other external doors

A wheelchair user should be able to easily collect post without the need to reach down to a low level. Where a letter box is provided, a letter cage should be fitted to the rear of the door. There should be enough space internally on the hinge side of the door, to maintain the clear opening width and prevent the cage hitting the adjacent wall when the door is fully open.

### Figure 4.5 Retrieving post

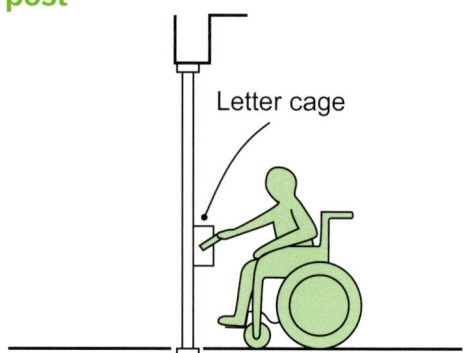

All entrance doors should have an accessible threshold as even minimal upstands can still create a barrier to movement and are difficult to propel a wheelchair over. The relationship between door thresholds and adjoining floor finishes should also be considered, so that upstands are not created when flooring is laid.

Doors into gardens and onto balconies and terraces should be carefully chosen to ensure that an accessible threshold is achievable. French doors, sliding and bi-fold doors in particular often have a raised threshold and may therefore need to be recessed. Specification of the ironmongery profile, as well as door weight and ease of movement, should be considered to ensure that excessive opening force is not required.

It can be difficult for a wheelchair user to operate double doors, as they need to hold one door open while opening the adjacent door. Where double doors are provided, it should be possible for a wheelchair user to move through a single leaf, without the need to also open the adjacent leaf for additional width.

Doors into gardens or onto balconies or terraces may open outwards but wind can make them difficult to hold open or to close. Inward opening doors may be a more practical solution, provided there is adequate space internally to negotiate the door swing without compromising living or circulation areas.

External doors should be sited so they do not inhibit the positioning of furniture.

### Figure 4.6 External doors

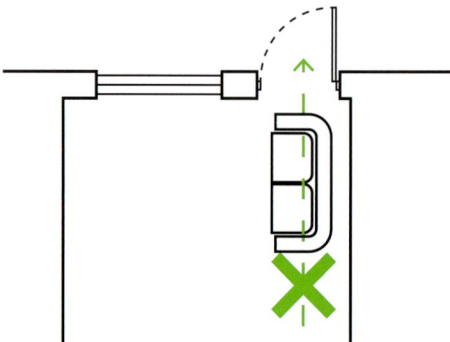

# Technical provisions

## 4.2 Technical provisions

*When designing to Part M of the Building Regulations, Approved Document M, Volume 1, M4(3) Category 3, the technical provisions detailed in this chapter are relevant to both M4(3)(2)(a) – **Wheelchair adaptable dwellings** and M4(3)(2)(b) – **Wheelchair accessible dwellings**.*

### 4.2.1 Principal private entrance door

The principal private entrance door is the entrance to an individual dwelling that a visitor not familiar with the dwelling would normally expect to approach. This is usually the 'front door' to a house or the private entrance to an individual dwelling within a block of flats.  3.22

A level landing with a minimum width and depth of 1500mm should be provided outside the principal private entrance, clear of any door swing or obstruction.  3.22a

Where the dwelling is within a block of flats, the corridor leading to the dwelling should be designed to accommodate the level landing of minimum width and depth of 1500mm in front of the flat entrance door.  3.22a

The landing should be covered for a minimum width and depth of 1200mm.  3.22b

**Figure 4.7 Landing and cover to the principal private entrance**  Diag 3.3

(a) 1200mm × 1200mm canopy
(b) 1500mm × 1500mm level landing clear of any door swing

Wheelchair Housing Design Guide

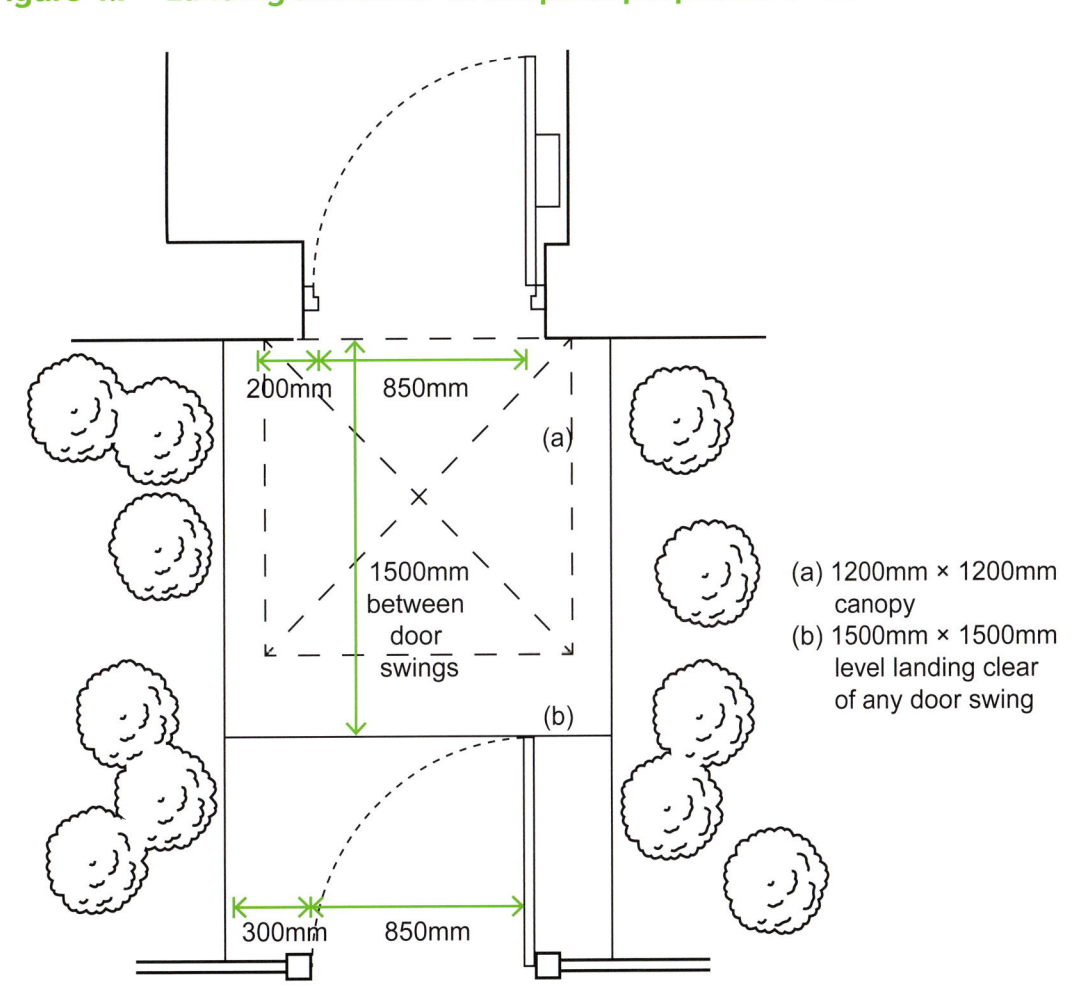

# 4 Technical provisions

| | AD M |
|---|---|
| Lighting should be provided that uses fully diffused luminaires and is activated by a dusk to dawn timer or by detecting motion. | 3.22c |
| The door entry control should be mounted 900mm to 1000mm above finished ground level and a minimum of 300mm away from any return corner. A door entry phone with remote door release facility should be provided in the main living space and in the principal bedroom. Any ironmongery, such as handles, locks, latches and catches, should be easy to grip and operate and fitted with the centre line 850mm to 1000mm above floor level. | 3.22k 3.44i 3.44f |

### Figure 4.8 Door entry controls

3.22k

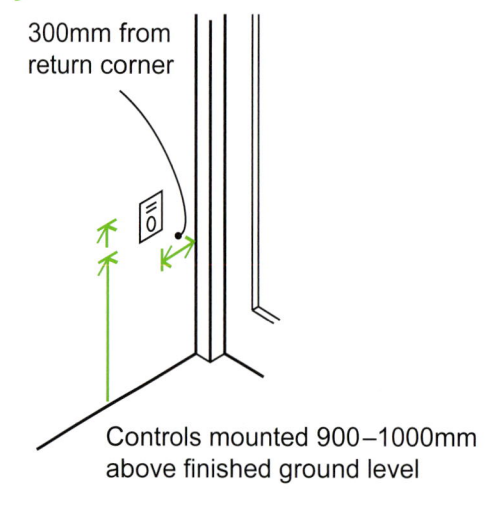

*300mm from return corner*

*Controls mounted 900–1000mm above finished ground level*

| | |
|---|---|
| There should be sufficient space alongside the door for a wheelchair user to approach and reach the handle: | |
| • Provide a minimum 300mm nib to the leading edge (pull side) of the door, with the extra width created by this nib maintained for a minimum distance of 1800mm beyond it. | 3.22e |
| • Provide a minimum 200mm nib to the following edge (push side) of the door, with the extra width created by this nib maintained for a minimum distance of 1500mm beyond it. | 3.22g |
| A minimum 150mm clear nib should also be provided to the hinge side of the door to allow for the fitting of a letter cage to the inside of the door. | 3.22e |
| The door should be located reasonably centrally within the thickness of the wall so that the depth of the reveal on the leading face of the door is no greater than 200mm. | 3.22h |
| The door should have an accessible threshold, i.e. the threshold should be level or, if raised, have a total height of no more than 15mm, a minimum number of upstands and slopes and any upstands higher than 5mm chamfered. The door should provide a minimum clear opening width of 850mm. Where there are double doors, the main (or leading) leaf should provide the required minimum clear opening width of 850mm. | 3.22i 3.22f 3.22g |
| A clear turning circle with a minimum 1500mm diameter should be provided inside the entrance area, in front of the door when closed. This turning circle should be clear of any localised obstructions and should not overlap the required wheelchair storage and transfer space (refer to Chapter 5: Dwelling circulation areas and storage) where this is positioned adjacent to the door. Where there is a lobby or porch, the doors should be a minimum of 1500mm apart and have a minimum of 1500mm between door swings. | 3.22d 3.22j |

Wheelchair Housing Design Guide

# Technical provisions

**4**

AD M
3.22l

Diag 3.3

A fused spur suitable for the fitting of a powered door opener should be provided on the hinge side of the door.

**Figure 4.9   Principal private entrance**

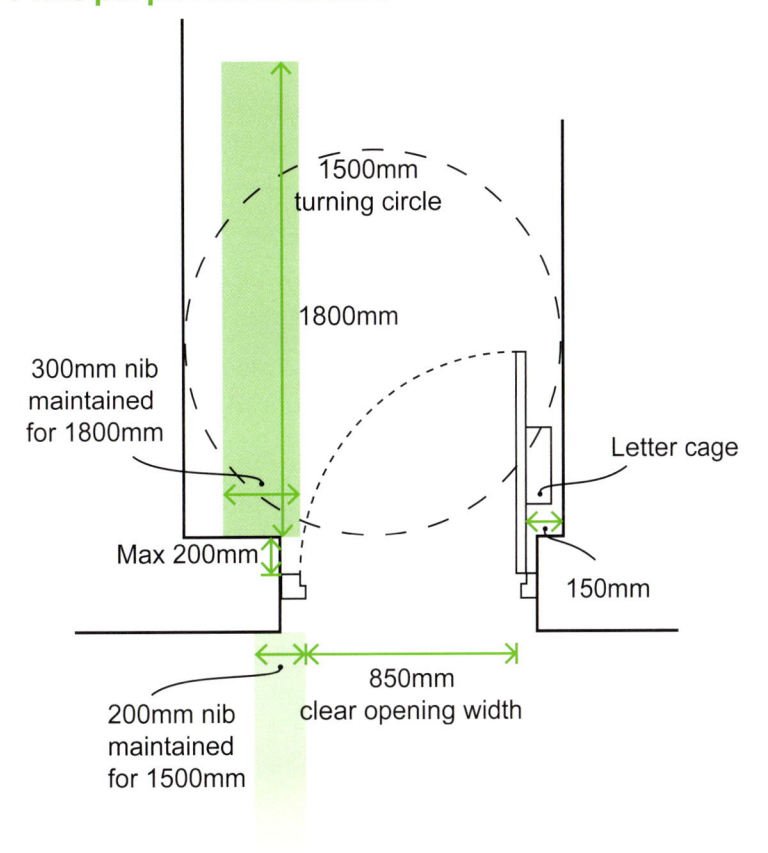

- 1500mm turning circle
- 1800mm
- 300mm nib maintained for 1800mm
- Max 200mm
- Letter cage
- 150mm
- 850mm clear opening width
- 200mm nib maintained for 1500mm

**Additional good practice recommendations**

- Where the private entrance door opens to the outside, the underside of the entrance cover should be a maximum height of 2300mm for effective weather protection, with coverage extending beyond the door on the lock side.

- Door entry controls should:
    - have an appropriately positioned camera, where video entry is included, that provides a good view of all callers
    - have call buttons that are easy to identify and operate with reduced dexterity and strength
    - include visual indicators which indicate response.

- Entrance matting, where provided, should be firm and recessed with no lip or upstand so the mat is level with the adjacent flooring. Provide flooring in the entrance area that is durable and easy to clean, and avoid coir matting or carpet in this area.

Wheelchair Housing Design Guide

# 4 Technical provisions

| | AD M |
|---|---|

### 4.2.2 Other private external doors

A wheelchair user should be able to easily negotiate all other private external doors, including those to and from a private garden, balcony, terrace, garage, carport, conservatory and any storage areas associated with the dwelling. — 3.23, 3.19

There should be sufficient space alongside the door for a wheelchair user to pull up and reach the handle:

- Provide a minimum 300mm nib to the leading edge (pull side) of the door, with the extra width created by this nib maintained for a minimum distance of 1800mm beyond it. — 3.23

    Note 1: Where sliding doors are used, both sides of the door should be treated as the leading edge.

    Note 2: Where a sliding door or outward opening door opens onto a balcony or terrace, the width of the leading edge nib is maintained for a distance of 1800mm, which will affect the overall depth of the balcony (refer to Figure 13.3).

- Provide a minimum 200mm nib to the following edge (push side) of the door, with the extra width created by this nib maintained for a minimum distance of 1500mm beyond it. — 3.23, 3.22g

The door should be located reasonably centrally within the thickness of the wall so that the depth of the reveal on the leading face of the door is no greater than 200mm. — 3.23, 3.22h

Doors should have an accessible threshold, i.e. the threshold should be level or, if raised, should have a total height of no more than 15mm, a minimum number of upstands and slopes and any upstands higher than 5mm chamfered. Doors should provide a minimum clear opening width of 850mm. Where there are double doors, the main (or leading) leaf should provide the required minimum clear opening width of 850mm. — 3.23, 3.22i, 3.22f, 3.22g

Where there is a lobby or porch the doors should be a minimum of 1500mm apart and have a minimum of 1500mm between the door swings. — 3.23, 3.22j

Door entry controls, where provided, should be located between 900mm and 1000mm above finished ground level and a minimum of 300mm away from any external return corner. Ironmongery, such as handles, locks, latches and catches, should be easy to grip and use and fitted 850mm to 1000mm above ground level. — 3.23, 3.22k, 3.44f

> **Additional good practice recommendations**
> - Provide fully diffused lighting to all private entrances that clearly illuminates any approach route, door handle and lock.

# Dwelling circulation areas and storage 5

## Principle

Circulation in dwellings should be designed to ensure that wheelchair users, and those assisting them, can approach and negotiate doors, hallways and circulation routes into and between rooms and storage areas.

## 5.1 Design considerations

The key to good circulation is to provide the appropriate space required to manoeuvre and turn a wheelchair easily. Layouts should also ensure there are no obstructions or projecting features that could restrict access or cause injury to a wheelchair user's hands, feet or elbows.

Wheelchair users may use a variety of manual and powered wheelchairs, potentially require assistance from another person, or have impairments which may impact on their spatial awareness and ability to manoeuver in tight spaces. Most wheelchair users, both independent and assisted, do not tend to move in a straight line and are unable to turn in a tight circle.

There are a number of critical wheelchair manoeuvres that are likely to be required within a dwelling:

- 90° and 180° turns within circulation routes
- 90° turns to approach and pass through doorways
- negotiation of a door swing when a door opens outwards into circulation space
- head-on approaches to doors opening against, or in, the direction of travel when the door is at the end of a corridor.

### Figure 5.1  Critical wheelchair manoeuvres

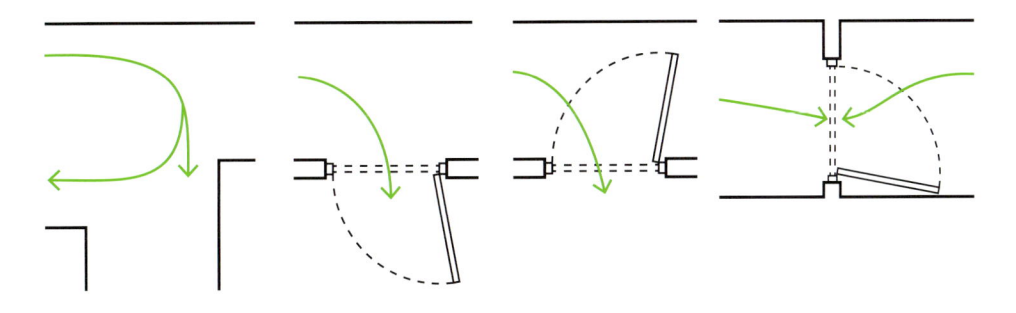

# 5 Dwelling circulation areas and storage

Providing sufficient space for these manoeuvres within a dwelling reduces the likelihood of damage to doors and walls and can avoid the need for additional protection, such as kickplates, which can have an institutional appearance.

**Circulation areas**

Corridors and other circulation areas should be wide enough to allow for critical wheelchair manoeuvres. It is easier for a wheelchair user to approach a door head-on than it is to enter a doorway at an angle. Where the approach is not head-on, a wider corridor provides additional space for this manoeuvre.

### Figure 5.2  Approaching a door at an angle

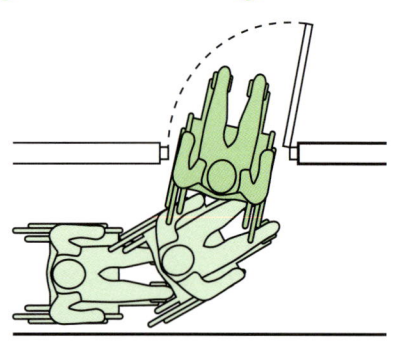

To improve circulation and ease of access into and between rooms, a large central hallway can be more effective than a long corridor design.

Long corridors, awkward turns into rooms and doorways positioned close to one another or to corners should be avoided. An outward opening door close to the corner of a circulation area/hallway and another inward opening doorway on the adjacent wall is difficult for a wheelchair user to negotiate.

### Figure 5.3  Doors in a corner

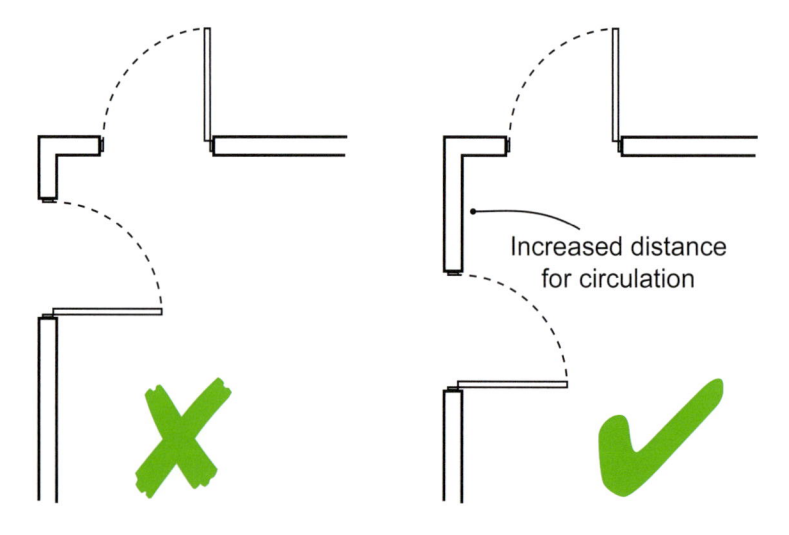

Increased distance for circulation

# Dwelling circulation areas and storage 5

Doors positioned opposite each other in a mirror image can reduce difficult manoeuvres.

**Figure 5.4   Doors opposite each other**

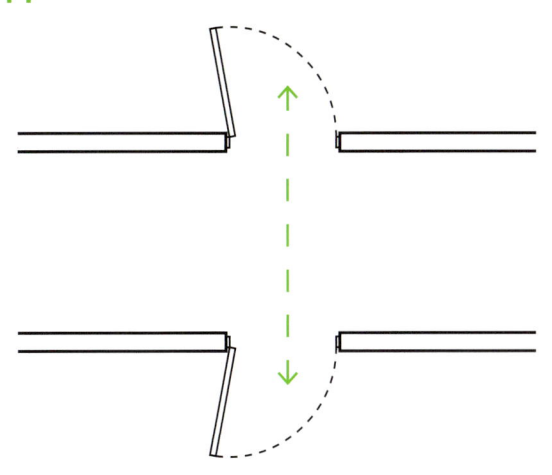

Consideration should also be given from the outset as to where radiators or any other potential obstructions will be positioned so that they do not impact on circulation, clear access zones or use of doorways.

**Figure 5.5   Localised obstructions**

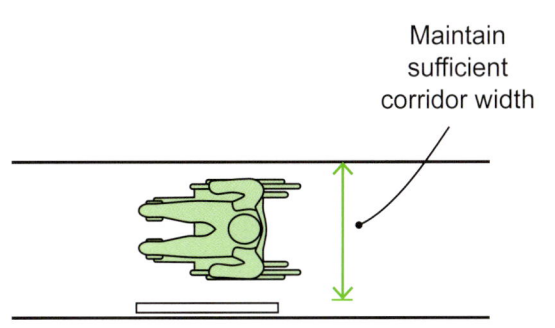

Flooring should facilitate ease of movement. Consideration should therefore be given to flooring which is slip-resistant, reduces trailing dirt and is easy to clean. There should be accessible thresholds between rooms and at cupboard doors to improve accessibility. Where there is a change in floor finishes, such as between a bathroom and bedroom, the finished floor levels should ensure an accessible threshold is achieved.

### Doors

All doors should have a sufficient clear opening width to enable a wheelchair user to easily move through and beyond the door. Where double doors are provided inside the dwelling, at least one leaf should provide sufficient width to manoeuvre through, as it is difficult to operate two doors simultaneously. If practicable, doors should open beyond 90° to assist entry and ensure handles do not impede access.

# 5 Dwelling circulation areas and storage

Additional space or clear door nibs are required on both sides of the door to enable a wheelchair user to get close enough to reach the handle and negotiate the door. To be effective, the additional width created by these nibs needs to be maintained for an adequate distance to enable a wheelchair user to manoeuvre or reverse and move through the door. Where doors are located at the end of corridors, this may require increased corridor width in front of the door to accommodate this extra circulation.

### Figure 5.6    Door nibs

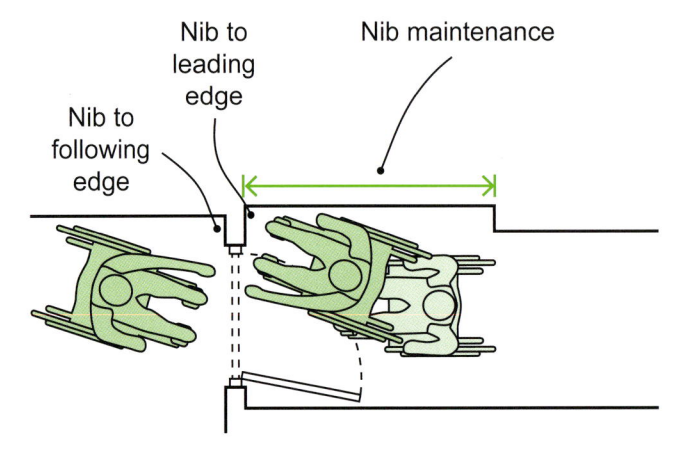

Sliding or pocket doors can assist where there is reduced circulation space on either side of the door. They should be robust and easy to operate, with no tracks or upstands in the floor. Where this type of door is fitted, adequate clear opening width (taking into account door handle projection) and sufficient space alongside the door to reach and use the handle on both sides is required. Both sides of a sliding door should be treated as a leading edge to ensure there is adequate space when approaching from either side. Specification of the ironmongery profile and height as well as door weight should be considered to ensure ease of use.

### Figure 5.7    Sliding/pocket doors

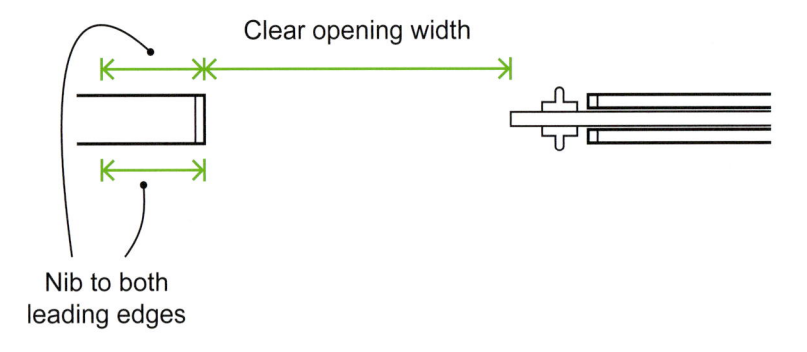

# Dwelling circulation areas and storage 5

### Wheelchair storage and transfer space

Wheelchair users often have more than one wheelchair, e.g. a powered wheelchair may be used outdoors and a manual wheelchair indoors.

Internally, there should be a dedicated space for a wheelchair user to transfer to a second wheelchair and store or charge the first wheelchair. This space should be located as close as possible to the principal entrance for ease of use when entering and leaving in order to avoid trailing dirt throughout the dwelling. It should also be clear of hallway/circulation routes and not impact on any approach zones or the space required to reach door handles or furniture.

A power socket should be provided within the space to enable charging and reduce the need to trail wires across a hallway or doorway.

### General built-in storage

Wheelchair users often require higher levels of storage to accommodate necessary items such as specialist equipment and medical supplies, which are usually delivered in large quantities.

Dedicated built-in storage space should be provided in addition to the wheelchair storage and transfer space. This storage space should be appropriate for the size of dwelling and number of bedrooms. Storage should be kept clear and not be used for other purposes, such as heat recovery units, boiler systems or washing machines.

The size, depth and height of storage and clear opening width of doors should provide optimal wheelchair access to internal storage space and shelving. Wide but shallow storage, approached sideways, can be a more efficient use of space as opposed to deep, narrow storage where it may be difficult to reach items stored at the back. Sliding doors can eliminate the extra space needed to avoid a door swing or can assist where space is restricted.

### Figure 5.8   Storage

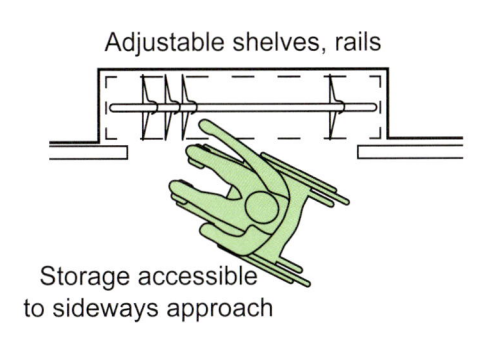

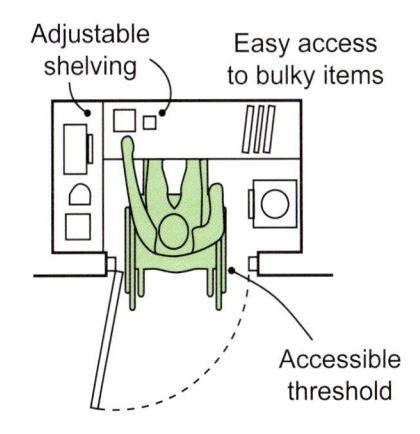

Wheelchair Housing Design Guide

# 5 Technical provisions

## 5.2 Technical provisions

*When designing to Part M of the Building Regulations, Approved Document M, Volume 1, M4(3) Category 3, the technical provisions detailed in this chapter are relevant to both M4(3)(2)(a) – **Wheelchair adaptable dwellings** and M4(3)(2)(b) – **Wheelchair accessible dwellings**. For wheelchair adaptable dwellings, paragraph 5.2.3 Wheelchair storage and transfer space should be read in conjunction with Chapter 14: Designing wheelchair adaptable dwellings, and the provisions outlined in paragraph 14.2.3.*

### 5.2.1 Circulation

| | AD M |
|---|---|
| To facilitate wheelchair movement into and between rooms, the clear width of every hallway, approach or landing should be a minimum of 1050mm. Where the approach to a doorway is not head-on, the minimum clear width of the hallway or approach should be 1200mm. | 3.24<br>3.24a<br>3.24b |

Any localised obstruction, such as a radiator, should not:

| | |
|---|---|
| • be longer than 2m in length | 3.24c |
| • reduce the minimum clear corridor width by more than 150mm | 3.24c |
| • be positioned opposite a doorway for a minimum distance of 1500mm centred on the door | Diag 3.4 |
| • be within 300mm of a change of direction in the hallway | Diag 3.4 |
| • be positioned within 1500mm centred on the doorway | Diag 3.4 |
| • be within the leading edge or following edge nib to any doorway and the minimum 1200mm distance the extra width created by the nib is maintained for beyond it | Diag 3.4 |
| • be positioned on the hinge side of the door (when approaching head-on) for a length of 800mm. | Diag 3.4 |

### Figure 5.9 Localised obstructions

Diag 3.4

An obstruction should not be positioned:
(a) opposite a doorway for a minimum distance of 1500mm centred on the doorway
(b) within 300mm of a change of direction in the hallway
(c) within 1500mm centred on the doorway
(d) within the leading edge or following edge nib to any doorway and the minimum 1200mm distance for which the extra width created by this nib is maintained beyond it
(e) on the hinge side of the door (when approaching head-on) for a length of 800mm

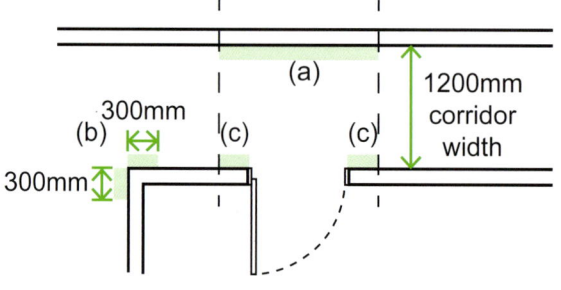

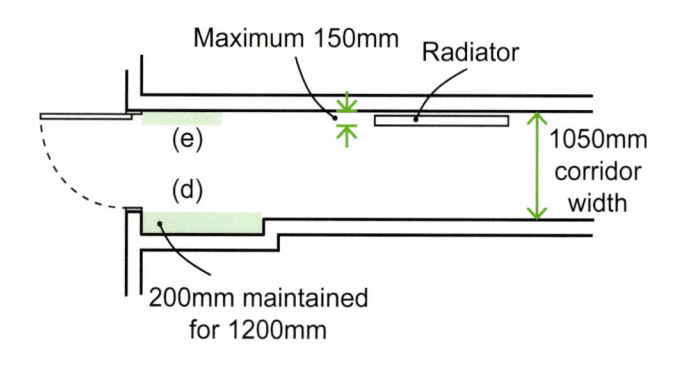

# Technical provisions 5

## AD M

Where an outward opening door is located close to a corner and another door is located on the return wall within 800mm of that corner, the leading edge of the outward opening door should be a minimum of 800mm from the corner. Where this is not provided, and the leading edge of the outward opening door is closer to the corner, the door on the return wall should be a minimum of 800mm from the corner. Alternatively, a minimum 1500mm turning circle should be positioned immediately outside the outward opening door to allow adequate manoeuvring space.

3.24e

Diag 3.5

Diag 3.5

### Figure 5.10   Outward opening doors close to a corner

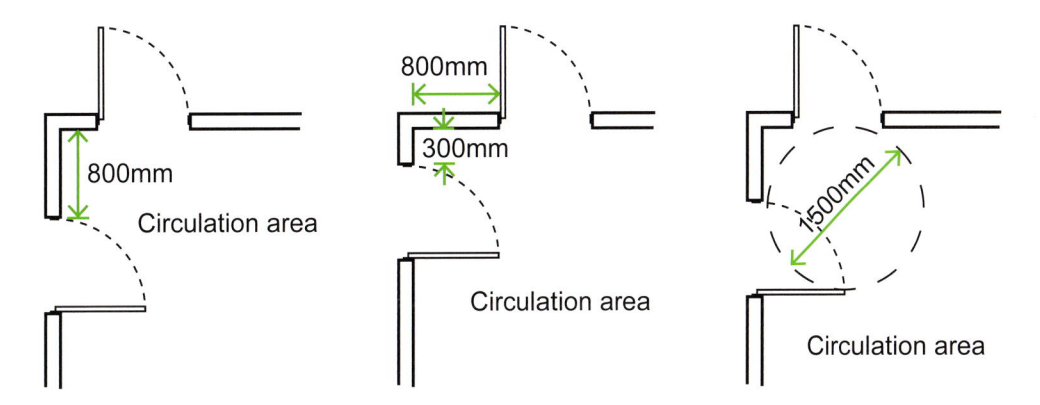

Diag 3.5

> **Additional good practice recommendations**
> - Where possible, provide a clear corridor width of 1200mm throughout the dwelling.

### 5.2.2   Doors (also refer to Chapter 11: Operating internal doors and windows)

To enable a wheelchair user to move easily into and between rooms, every door, whether hinged, sliding or pocket, should have a minimum 850mm clear opening width, irrespective of the direction of entry. This includes walk-in cupboards and storage.

3.24d
3.24 Note 1

### Figure 5.11   Clear opening width

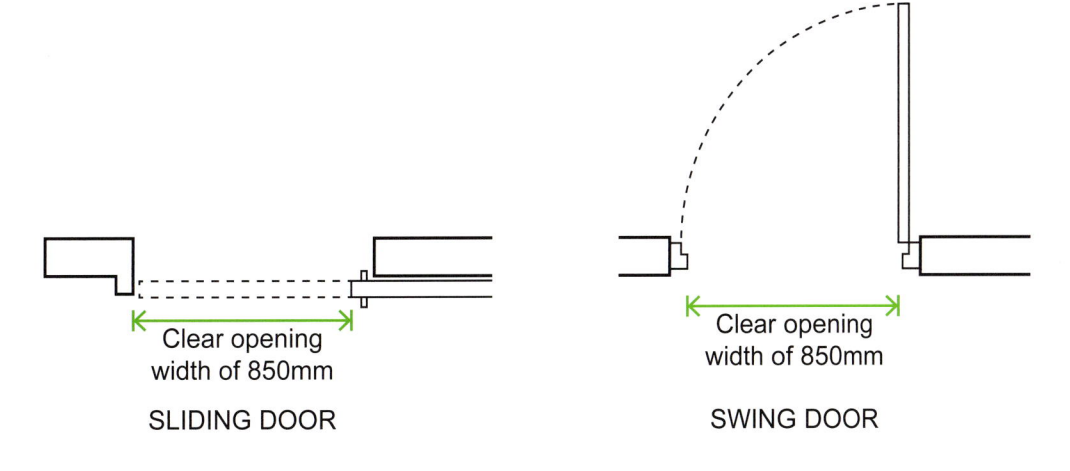

Diag 3.2

SLIDING DOOR — Clear opening width of 850mm

SWING DOOR — Clear opening width of 850mm

Wheelchair Housing Design Guide

# 5 Technical provisions

| | AD M |
|---|---|

Ensure that there is space alongside the door to enable a wheelchair user to pull up and reach the handle by providing:

- A minimum 300mm nib to the leading edge (pull side) of the door, with the extra width created by this nib maintained for a minimum distance of 1200mm beyond it. — 3.24f, Diag 3.4

    Note 1: Where sliding doors are used, both sides of the door should be treated as the leading edge.

- A minimum 200mm nib to the following edge (push side) of the door, with the extra width created by this nib maintained for a minimum distance of 1200mm beyond it. — 3.24g, Diag 3.4

Note: Double doors effectively provide nibs where each leaf is a minimum of 300mm wide. — 3.24 Note 2

Where doors are located at the end of corridors, the corridor may need to be wider than 1050mm to accommodate the clear opening width and door nib. The extra width created by this nib also needs to be maintained for a distance of 1200mm beyond it. — 3.24d, 3.24f, 3.24g, Diag 3.4

The provisions of 5.2.1 and 5.2.2 do not apply to cupboards unless they are large enough to be entered or to en-suite bathroom facilities that are additional to those required by Table H and I. — 3.24 Note 1

### 5.2.3 Wheelchair storage and transfer space

The dwelling should have a suitable storage and transfer space to enable a wheelchair user to charge and store up to two wheelchairs and transfer between an outdoor and an indoor wheelchair. Provide a wheelchair storage and transfer space a minimum 1100mm deep and 1700mm wide. This should be accessible from a space with a minimum clear width of 1200mm, positioned adjacent to one of the long sides. This wheelchair storage and transfer space is in addition to all other requirements, such as general built-in storage, clear access zones, clear nibs to doorways and minimum floor areas for bedroom and living areas. — 3.25, 3.25a, 3.25b, 3.25d

Note: To enable a standing transfer between two wheelchairs, the wheelchair storage and transfer space should have full head height.

The dedicated space should be available on the entrance storey of a dwelling and preferably close to the principal private entrance. — 3.25a

A power socket should be provided within the wheelchair storage and transfer space, with its centre line 700mm to 1000mm above floor level and a minimum of 700mm (measured horizontally) from any inside corner. — 3.25c, 3.44b

# Technical provisions

## Figure 5.12  Wheelchair storage and transfer space

Centre line of socket a minimum of 700mm from corner

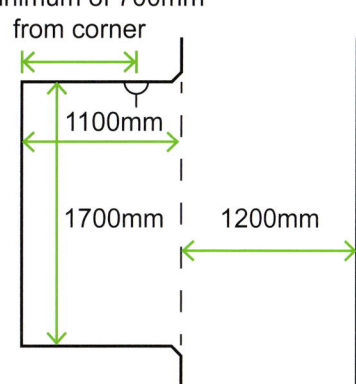

1100mm

1700mm   1200mm

> **Additional good practice recommendations**
> - Provide a double power socket to allow the charging of two wheelchairs simultaneously.

### 5.2.4  General built-in storage

General built-in storage should be provided in accordance with Table B. This should be in addition to all other spatial and furniture requirements, e.g. bedroom wardrobes. Refer to Chapter 9: Bedrooms and Appendix 1: Furniture schedule.

Storage areas with headroom below 900mm should not be included as part of the overall required storage provision. Storage areas with reduced headroom between 900mm and 1500mm are only measured at 50% of the area. The full area under a stair that forms part of the storage provision should only be counted as 1m².

| Table B | Minimum area of general built-in storage | | | | | |
|---|---|---|---|---|---|---|
| **Number of bedrooms** | 1 | 2 | 3 | 4 | 5 | 6 |
| Minimum storage area (m²) | 1.5 | 2.0 | 2.5 | 3.0 | 3.5 | 4.0 |

Note: General built-in storage is calculated on the number of bedrooms, and not bedspaces.

> **Additional good practice recommendations**
> - In addition to the minimum area of general built-in storage required in Table B, provide an additional 0.5m² per dwelling, to accommodate additional medical supplies and equipment.

**AD M**
Diag 3.6

3.26
Appx D

3.26 Note

Table 3.1

Wheelchair Housing Design Guide

# Moving between levels within the dwelling

## 6

## Principle

Where individual dwellings are designed on more than one storey, there should be wheelchair access to all storeys within the dwelling.

## 6.1 Design considerations

Single storey dwellings provide all accommodation on a single level without reliance on a private internal lift. Wheelchair accessible dwellings over more than one floor will require a commissioned lift from the outset to ensure access to the entire dwelling. The inclusion of a lift has significant implications for the overall footprint of a dwelling, capital and running costs, convenience, servicing and maintenance.

**Lift provision**

A vertical lift enables a wheelchair user to move between storeys in their home while seated in their wheelchair. In order for a wheelchair user to access all parts of the dwelling, the lift should connect circulation areas on each floor level without compromising privacy or impacting on circulation and living space.

Lifts meeting minimum standards are typically only large enough, or designed for, an independent wheelchair user. In many instances however, it is necessary or desirable for a wheelchair user to be accompanied, e.g. a parent may need to travel in the lift with a disabled child. The lift size and specification should therefore aim to accommodate accompanied wheelchair users as well as a range of wheelchair sizes, including larger powered or tilt-in-space wheelchairs.

There are various types of lifts available. Through-floor lifts have a partially enclosed lifting platform and move between two storeys. In dwellings of more than two storeys, a fully enclosed liftway is likely to be required. The number of storeys within a home will therefore determine which type of lift is the most suitable for the dwelling.

Through-floor lifts can travel between no more than two storeys and have certain features which should be taken into account when incorporating them into the design of a dwelling:

- The door can only be hinged to open consistently to either the left or right on both floors and generally does not open beyond 90° The dwelling design should incorporate space for the door swing to ensure there is appropriate circulation space and manoeuvrability at each level.

- In addition to internal controls, external controls can be wall mounted and hard wired at each floor level and/or remote controlled. This can enable a personal assistant to control the lift from outside where needed, or allow a wheelchair user to send the lift up or down to the other floor to free up circulation space.

- There may be manufacturer restrictions or insufficient space which prevent a standing person from accompanying a wheelchair user.

# 6 Moving between levels within the dwelling

When a lift is required to travel more than two storeys, different considerations will apply to the lifting platform:

- A larger overall footprint is required to accommodate a larger, enclosed lift.
- A shallow lift pit may be required on the lowest level to enable level access to the lift. Ramped access is not suitable as it is more difficult to negotiate and impacts on dwelling circulation space.
- Where feasible, controls should not require continual pressure during use as this can be difficult for some users.

To accommodate the lifting mechanics, the rear of the lift is located against a wall. The orientation of the lift will be the same at every level, with the door on the same narrower end. The overall layout of the dwelling therefore requires careful consideration to ensure that this lift position, and suitable clear circulation space in front of the lift at every level, can be achieved.

A wheelchair user will enter and exit the lift facing in the same direction, as there is insufficient space to turn within the lift. A powered door opener/closer with controls that are easy to reach and operate will increase independence and manoeuvrability.

For safety reasons, the lift door opening should be located away from the top of the stairs, and ideally be clear of the stair landing. There should be sufficient space to manoeuvre a safe distance away from the top step to reduce the risk of falling down the stairs.

**Private stairs**

To accommodate a range of users and scenarios, stairs should be easy to negotiate and enable an ambulant disabled person to access all levels within the dwelling. The stair design should also allow a stair-lift to be fitted between the entrance storey and the principal bedroom and bathroom on either the storey above or below. Straight flights of stairs are preferable, or where there is change of direction, a full intermediate landing should be provided. Winder treads should be avoided as these are often difficult to use and do not always enable a stair-lift to be easily installed.

It should be noted that a stair-lift is not the same level of provision as a vertical lift. A stair-lift is typically used by ambulant disabled people, who are able to transfer on and off and to sit safely on the stair-lift while it ascends and descends.

# Technical provisions 6

## 6.2 Technical provisions

*When designing to Part M of the Building Regulations, Approved Document M, Volume 1, M4(3) Category 3, the technical provisions detailed in this chapter are relevant to both M4(3)(2)(a) – **Wheelchair adaptable dwellings** and M4(3)(2)(b) – **Wheelchair accessible dwellings**, with the exception of paragraph 6.2.1 Dwelling lift requirements. For wheelchair adaptable dwellings and the provisions relevant to dwelling lift requirements, refer to Chapter 14: Designing wheelchair adaptable dwellings, and the provisions outlined in paragraph 14.2.4.*

### 6.2.1 Dwelling lift requirements

Where a dwelling has more than one floor level, a wheelchair user should have access to all parts of the dwelling. In wheelchair accessible dwellings, a suitable through-floor lift or lifting platform should be installed and commissioned.

Perf – b
3.27
3.29

A continuous liftway is required which internally links the circulation areas on every floor level of the dwelling, irrespective of the number of storeys. The space required for suitable lift provision should be in addition to any other spatial requirements and clear access zones, and in particular should not be included in the minimum living, kitchen and dining floor areas set out in Chapter 7, Table C.

3.29a
3.29e
3.28e

The liftway should be a minimum of 1100mm wide and 1650mm long internally. The lift should be entered from the same one of its narrower ends, with the opposite end located against a wall, at every floor level.

3.29a
3.29b
3.29f

Lift doors should be power operated.

3.29g

A power socket, suitable for powering the lift, should be provided in close proximity to the liftway.

3.29d

There should be a minimum 1500mm clear turning circle in front of the liftway door at every floor level, clear of the liftway door when open at 90°. The turning circle should be clear of the top step of any adjacent flight of stairs.

3.29c
Diag 3.7

### Figure 6.1    Lift provision

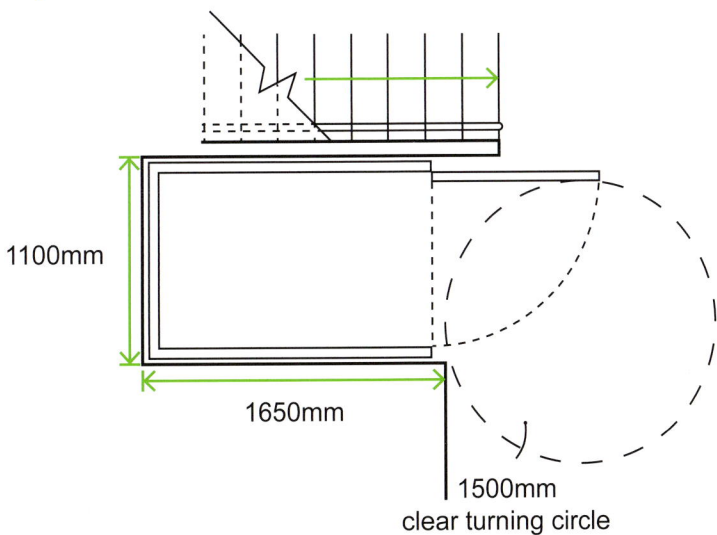

Diag 3.7

# 6  Technical provisions

In two storey dwellings provide a lift to BS 5900 or a lifting platform to BS EN 81-41. A lifting platform may require a larger liftway than stated in 6.2.1 and may also require a three-phase power supply. A lifting platform is likely to be needed when travelling more than two storeys.

**AD M**
3.29 Note

> **Additional good practice recommendations**
> - Where practicable, provide a larger lift than the minimum required in 6.2.1 to allow for a range of wheelchair sizes and accompanied wheelchair users in line with current lift standards.
> - The turning circle in front of the lift door should be clear of the landing at the top of the stairs.

### 6.2.2  Private stairs

Stairs should enable an ambulant disabled person to move within, and between, storeys. It should also be possible to fit a stair-lift to the stairs from the entrance storey to the principal bedroom and bathroom (as required by 10.2.5 or 10.2.6) on either the storey above or below.

3.30
3.41b

There should be no changes of level on any storey and access to all rooms and facilities within each storey should be step-free.

3.30a
3.30b
Perf – b

All private stairs should be a minimum of 850mm wide when measured 450mm above the pitch line of the treads (ignoring any newel post), and should adhere to the provisions of Part K[1] for private stairs.

3.30c
3.30e

### Figure 6.2  Stair width

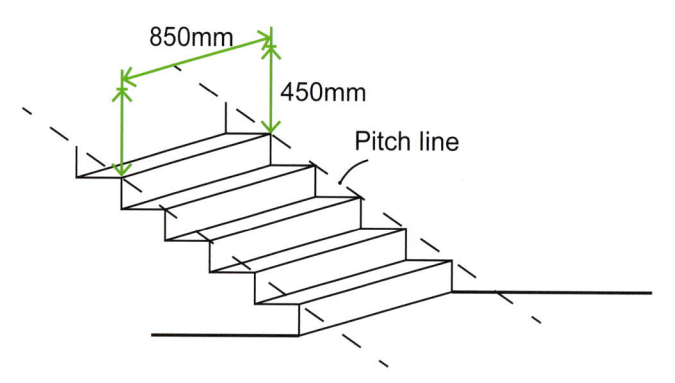

3.30c

A power socket, suitable for powering a stair-lift, should be provided close to the foot or head of the stairs between the entrance storey and the storey above or below containing the principal bedroom and bathroom required by 10.2.5 or 10.2.6.

3.30d
3.30
3.41b

> **Additional good practice recommendations**
> - Winder treads should be avoided.

---

[1]  Refer to Approved Document K.

# 7 Using living spaces

## Principle

Living spaces should be designed to enable a wheelchair user to socialise with friends and family. These areas should be suitably sized to accommodate a range of furniture in addition to adequate circulation space to reach and use all doors, windows and furniture.

## 7.1 Design considerations

For ease of use, living areas should be on the entrance storey of the dwelling. There should be a convenient relationship between the kitchen, dining and living room, enabling wheelchair users to move easily between these areas and carry food and drinks.

Open plan living can provide a functional and space efficient layout; however, people may prefer a separate kitchen for cooking, safety and noise reduction purposes. In larger homes, a living room separate from the kitchen/kitchen diner or provision of a second living space is preferred.

Careful consideration should be given to the positioning of windows, doorways, full height glazing, radiators and socket outlets as these can impact on where furniture is placed in a room. Design layouts should enable functional use of the furniture and provide sufficient overall space.

Living rooms which are long and narrow, have recessed areas or which taper into an angle are generally unsuitable. Living and dining areas should be large enough to accommodate a standard range of furniture while providing clear circulation space and access zones. Some wheelchair users find a perimeter layout of furniture, where furniture is moved back against outer walls, facilitates manoeuvring and maximises available circulation space.

Sufficient space should be available to transfer from a wheelchair to a chair and to approach and use furniture, such as a dining table or bookshelf. Consideration may need to be given to the future installation of an overhead hoist to assist with transfers, dependent on individual need.

A remote entrance door release function should be provided in the living space to assist wheelchair users who may find it more difficult and time consuming to leave the room to open the entrance door to visitors.

Furniture positioning should allow a clear access route to any external doors, such as those to a balcony or garden.

Where glazed doors provide the only external opening to living areas, wheelchair users on the ground floor may not feel secure leaving these open for ventilation. In this instance, an opening window with a clear access route should be provided.

Wheelchair Housing Design Guide

# 7 Using living spaces

**Figure 7.1    Example kitchen, dining and living areas**

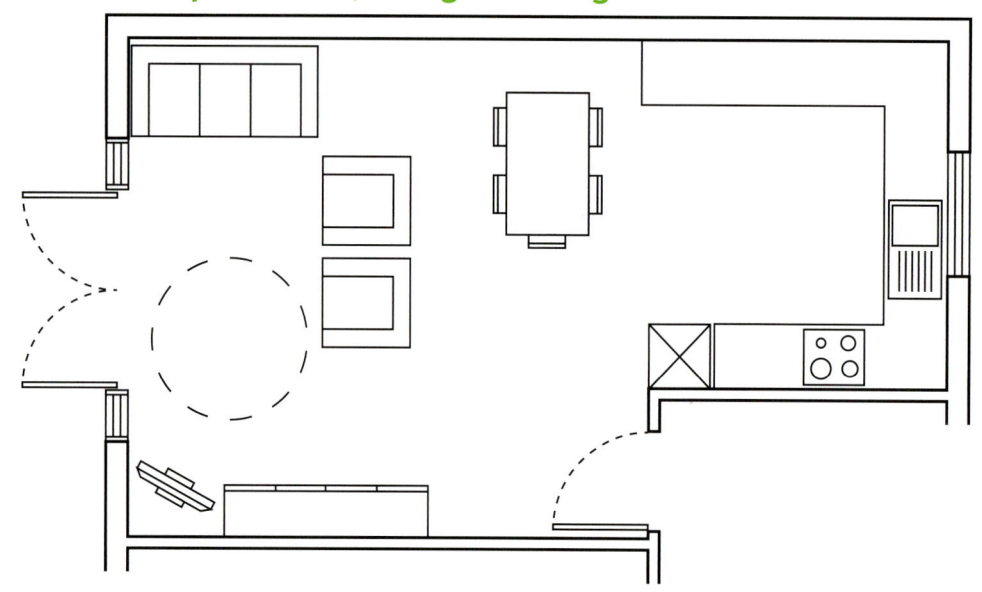

Unobstructed views to the outside from a seated position are particularly important in wheelchair accessible dwellings where a person may spend more time indoors, and can contribute to a sense of wellbeing. Window glazing should be of a suitable height to allow views out and transoms, mullions and other obstructions should be carefully positioned so they do not interrupt sight lines. Window handles should be at an accessible height to allow wheelchair users to control ventilation independently.

**Figure 7.2    Window glazing**

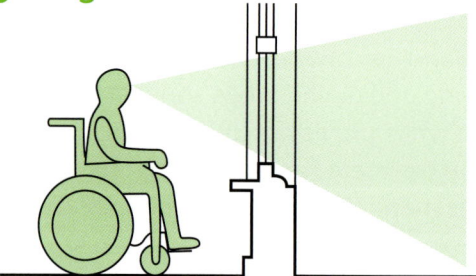

Bay windows or rooms with dual aspect increase the range of views and can improve natural light levels.

**Figure 7.3    Dual aspect windows**

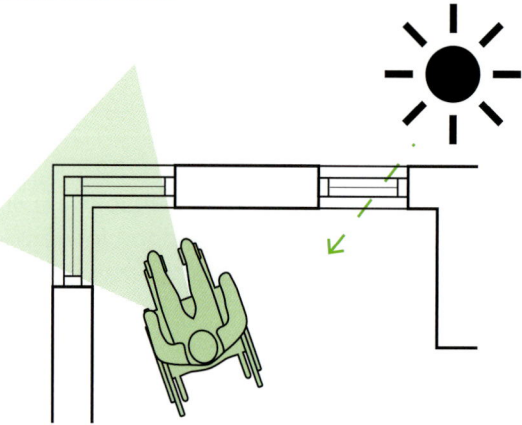

Wheelchair Housing Design Guide

# Technical provisions

## 7.2 Technical provisions

*When designing to Part M of the Building Regulations, Approved Document M, Volume 1, M4(3) Category 3, the technical provisions detailed in this chapter are relevant to both M4(3)(2)(a) – **Wheelchair adaptable dwellings** and M4(3)(2)(b) – **Wheelchair accessible dwellings**.*

To demonstrate that a dwelling is capable of meeting the functional and spatial provisions for a wheelchair user dwelling, furnished plan layouts that show the necessary clear access zones, other provisions and the furniture of the furniture schedule included in Appendix 1, should be shown at a scale of at least 1:100.

*AD M*

3.20

> **Additional good practice recommendations**
> - Provide fully furnished plans at a scale of 1:50.

### 7.2.1 Living areas

The kitchen, principal living and principal eating area should be on the entrance storey with a convenient relationship between these areas, the WC and principal private entrance.

3.31
3.31a
3.32

Access to these living areas should be step-free.

3.30a
Perf – b

The kitchen and principal eating area should either be within the same room or connected to each other.

3.32a

### 7.2.2 Room size and layout

Provide the minimum combined internal floor area for living, dining and kitchen space for the number of bedspaces within the dwelling in accordance with Table C.

3.31b
Table 3.2

| Table C | Minimum combined floor area for living, dining and kitchen space |||||||
|---|---|---|---|---|---|---|---|
| **Number of bedspaces*** | 2 | 3 | 4 | 5 | 6 | 7 | 8 |
| Minimum floor area (m²) | 25 | 27 | 29 | 31 | 33 | 35 | 37 |

Table 3.2

*\* For the purposes of establishing the number of bedspaces relevant to these requirements, a bedroom at or above 8.5m² and below 12.5m² is counted as one bedspace, and equal to or greater than 12.5m² as two bedspaces.*

3.36 Note 3

> **Additional good practice recommendations**
> - The principal living area should have a minimum width of 4000mm. Rooms should be a regular shape without awkward angles or corners.

### 7.2.3 Clear access zones and furniture provision

Provide a clear turning circle of minimum diameter 1500mm, or a turning ellipse of 1700mm × 1400mm in the living and dining area of every dwelling.

Appx D

Wheelchair Housing Design Guide

# 7 Technical provisions

Provide minimum furniture requirements for living and dining spaces, based on number of bedspaces, in accordance with Table D.

| Table D | Minimum furniture requirements for living and dining space | | | | | | | |
|---|---|---|---|---|---|---|---|---|
| | | **Number of bedspaces*** | | | | | | |
| | | 2 | 3 | 4 | 5 | 6 | 7 | 8[2] |
| **Furniture to be shown** | **Furniture size (mm)** | **Number of furniture items required** | | | | | | |
| Arm chair (or number of sofa seats in addition to minimum sofa provision) | 850 × 850 | 2 | 3 | 1 | 2 | 3 | 4 | 5 |
| 2 seat settee (optional) | 850 × 1300 | | | | | | | |
| 3 seat settee | 850 × 1850 | | | 1 | 1 | 1 | 1 | 1 |
| TV | 220 × 650 | 1 | 1 | 1 | 1 | 1 | 1 | 1 |
| Storage units | 500 × length shown (1 only required) | 1000 | 1000 | 1500 | 2000 | 2000 | 2000 | 2000 |
| Dining table | 800 × length shown (1 only required) | 800 | 1000 | 1200 | 1350 | 1500 | 1650 | 1800 |
| Dining chair | | 2 | 3 | 4 | 5 | 6 | 7 | 8 |

\* For the purposes of establishing the number of bedspaces relevant to these requirements, a bedroom at or above 8.5m² and below 12.5m² is counted as one bedspace, and equal to or greater than 12.5m² as two bedspaces.

AD M
3.20

Appx D

3.36 Note 3

---

[2] Furniture requirement for eight bedspace dwellings is assumed based on Approved Document M, Volume 1 (2015 edition incorporating 2016 amendments) requirements for other dwelling sizes.

# Technical provisions

> **Additional good practice recommendations**
> 
> - Ensure the positioning of radiators does not inhibit a reasonable furniture layout or wheelchair manoeuvrability.
> 
> - Provide a 1000mm clear space to transfer to at least one living room chair and 1000mm clear approach route to the dining table.
> 
> **Figure 7.4   Access to furniture**
> 
> Space to transfer to chair (1000mm)
> 
> Space to approach tables with leg space under (1000mm)

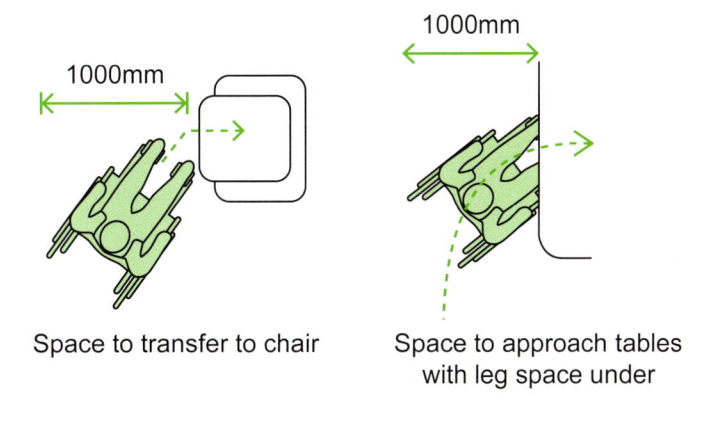

## 7.2.4   Windows

Ensure glazing to the principal window in the principal living area starts a maximum of 850mm above floor level.[3] — 3.31c

At least one window in the principal living area should have the handle 700mm to 1000mm above floor level, unless fitted with a remote opening device within the same height range. All other windows should have the handle 450mm to 1200mm above floor level unless fitted with remote opening devices within the same height range. — 3.44d / 3.44e

**Figure 7.5   Window handle height in the principal living area** — 3.44d

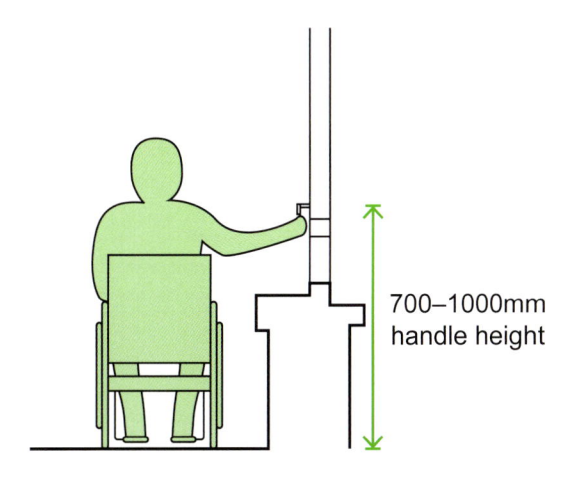

700–1000mm handle height

---

[3] Approved Document M, Volume 1 (2015 edition incorporating 2016 amendments) indicates glazing to this area should start a maximum of 850mm above floor level or at the minimum height reasonable to achieve compliance with Approved Document K for guarding to windows. — 3.31c

Wheelchair Housing Design Guide

# 7 Technical provisions

**AD M**

> **Additional good practice recommendations**
> 
> - Provide at least one window in the living area that can be opened for ventilation. There should be a minimum 750mm wide access route to this window, clear of all furniture.

### 7.2.5 Door entry phone

A door entry phone with remote door release facility should be provided in the main living space (and bedroom). The door entry phone should be located with the centre line of any control button(s) or areas of a touch screen required for operation 700mm to 1000mm above floor level and a minimum of 700mm (measured horizontally) from any inside corner. Refer to Chapter 12: Services and controls.

3.44i
3.44b

# Using the kitchen 8

## Principle

The layout and design of a wheelchair accessible kitchen is important to maximise independence and health and wellbeing for a disabled person, their family and personal assistants. The specification of the units, cooking facilities and the space available requires careful planning to ensure the kitchen is functional and can be easily and safely accessed.

## 8.1 Design considerations

The kitchen should be on the entrance storey, with a convenient relationship between the dining area and living space. The kitchen and dining areas should be combined or directly connected, to enable a wheelchair user to easily carry food and drinks between the two.

**Layouts**

The kitchen should be designed to incorporate a continuous section of worktop, appliances and units. Careful consideration should be given to the positioning of doors and windows so as to optimise the continuous working area and circulation space. Adequate full height wall space should be provided to accommodate wall units, tall units, a hob and extractor fan and appliances, including the oven housing, fridge, freezer and storage.

The kitchen may be used by a wheelchair user, family members or personal assistants and should be designed to provide flexibility for concurrent use by more than one person. The layout should maximise the range of operations possible from a single position, to enable a wheelchair user to carry out activities without the need for excessive manoeuvring or repositioning. Sufficient space should also be available for a wheelchair user to turn and manoeuvre between appliances and units.

An effective layout enables:

- ease of approach and use
- a convenient relationship between essential activities
- avoidance of disruptive sequencing.

# 8 Using the kitchen

### Figure 8.1  Sequence of activities

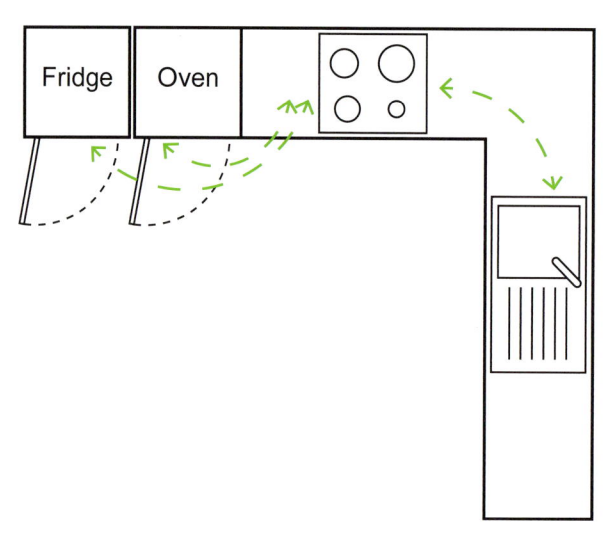

The oven, hob, sink and adequate space for food preparation should be located on the same run of worktop to create a safe and useable working area. A straight, L-shaped or U-shaped run enables a wheelchair user to slide or push a hot dish from the oven and hob to the sink. A wheelchair user should not have to lift or carry hot or heavy items from one side of the kitchen to another across an open space or across a doorway.

Any doors to and from the kitchen should not interrupt the continuous worktop run which houses the oven, hob and sink. The layout should avoid compromising working areas with cross routes.

### Figure 8.2  Kitchen layout

58  Wheelchair Housing Design Guide

# Using the kitchen  8

**Height adjustable worktop**

The hob, sink and food preparation area should be provided on a single section of height adjustable worktop with clear leg space underneath.

### Figure 8.3   Height adjustable worktop

Height adjustable worktop with clear leg space underneath

Height adjustable worktop with clear leg space underneath

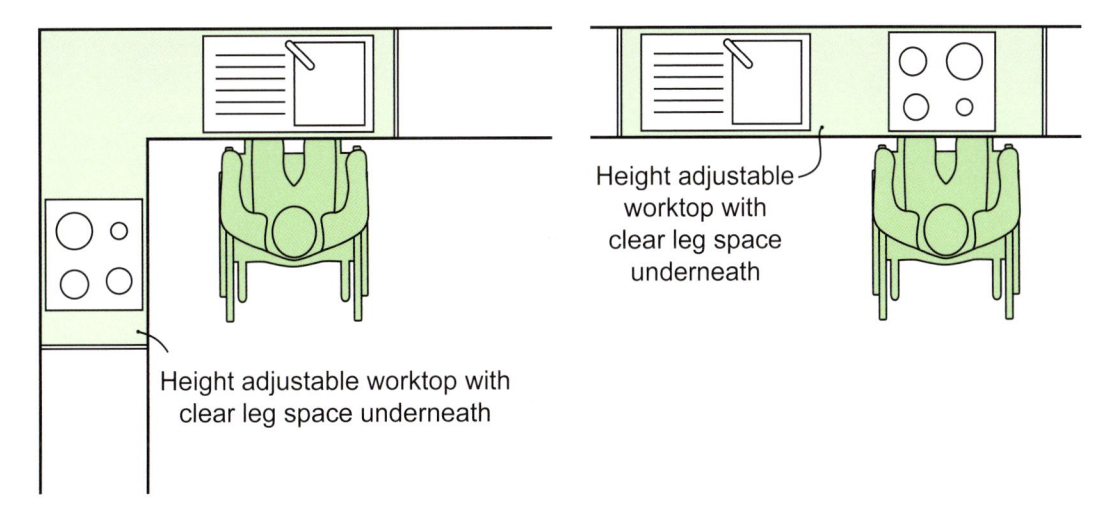

The height adjustable worktop brackets should not have diagonal struts that compromise the clear leg space below. This clear space should not be used for storage of appliances, bins or moveable furniture.

### Figure 8.4   Worktop fixings

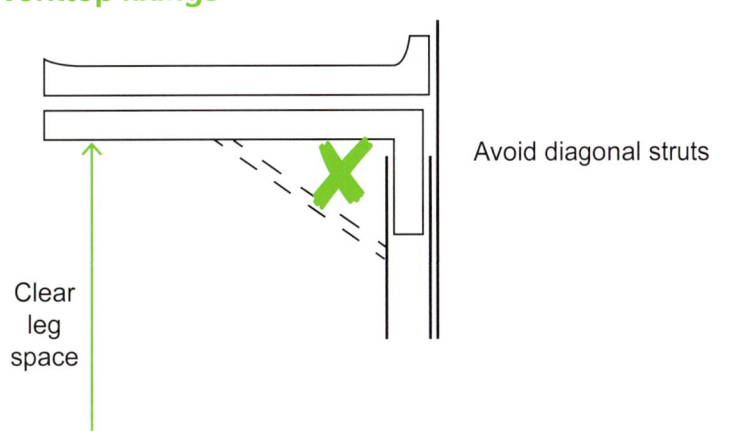

Avoid diagonal struts

Clear leg space

The height adjustable section could either be provided as a rise and fall worktop or as a fixed worktop, capable of being easily re-fixed at alternative heights. Early consideration should be given to this choice, as an additional depth of worktop may be required for the positioning of services, so that service outlets do not compromise clear leg space under the worktop. Outlets from the sink, washing machine and dishwasher should drain effectively with the worktop in a lowered position.

A wheelchair user should be able to wheel forward far enough to reach and use the worktop, sink and wall sockets. If a front fascia is provided, this should not compromise the height of the clear leg space.

Wheelchair Housing Design Guide

# 8 Using the kitchen

**Figure 8.5   Clear leg space**

### Sink

The sink should be shallow enough to provide leg space underneath. Waste traps should preferably be to the rear of the bowl and the underside of the sink should be insulated to protect the wheelchair user's legs. All exposed plumbing and pipework should be covered by a removable access panel or door to allow for maintenance, without compromising the depth of clear leg space.

**Figure 8.6   Sink**

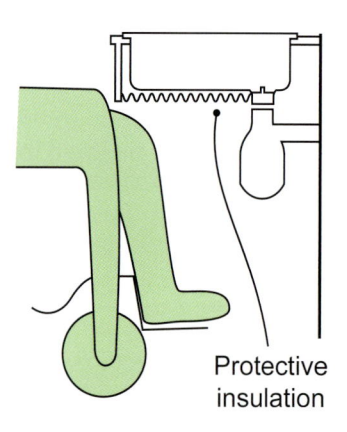

Protective insulation

A mixer tap with lever handle(s) should be provided to allow easy control of water temperature, flow and direction. A swivel arm tap of sufficient height and radius, such as a 'swan neck' type, allows the filling of a pot, pan or kettle on the adjacent drainer or work surface. Tap controls can be mounted on the front fascia of the worktop for easy reach.

### Kitchen units and storage

Storage provision is often insufficient in a wheelchair accessible kitchen as it is reduced when base units are omitted to provide clear leg space under the sink, hob and preparation areas. Sufficient storage space should therefore be provided to accommodate items such as food supplies, pans, crockery, cutlery and utensils, in addition to integral space provided for accessible refuse and recycling.

A wheelchair user may not be able to reach storage that is too high or too low. Where storage is likely to be beyond easy reach, an accessible solution could be incorporated. Accessible solutions may include internal pull out/down shelves or baskets, carousels and pull out larders. Alternatively, this out-of-reach storage could be used for items that do not require frequent use or which could be accessed by other household members. Tall wall units, while not fully accessible, increase the overall storage capacity of the kitchen.

# Using the kitchen    8

Moveable base units on wheels do not provide suitable storage as they can be heavy when full and are likely to be difficult to use and manoeuvre.

**Figure 8.7    Accessible storage solutions**

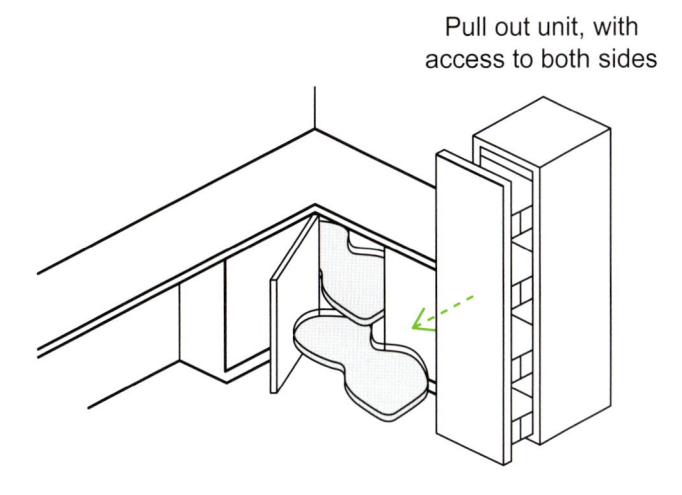

Pull out unit, with access to both sides

The provision of wall units and storage should be maximised where possible. While the upper shelves of standard wall units may be inaccessible to many wheelchair users, there are options that can be considered to improve access:

- wall units can be fitted at a lower height
- wall units can be fitted with robust pull down baskets
- electric rise and fall wall units can be provided.

When considering base unit provision, the inclusion of drawers rather than a hinged base unit door, can reduce the need to bend and reach into the back of the cupboard.

Cupboard and drawer handles should be easy to grip for a person with reduced dexterity, have no sharp edges and be positioned to maximise ease of use.

When fitting out a wheelchair accessible kitchen, kitchen units should be robust and durable enough to withstand the extra knocks and scrapes that can occur while manoeuvring a wheelchair. Plinths under base units can also reduce dirt accumulation in spaces where a wheelchair user may not be able to clean effectively.

### Appliances

The choice of appliances and their location can impact the overall kitchen usability and should be considered from the outset. The direction in which the appliance doors open can also affect the ease of access and use and the ability to transfer items.

Freestanding appliances can be deeper than integrated appliances. This should be taken into account when specifying the depth of the worktop or allocated space to avoid appliances protruding beyond the edge of the worktop and causing an obstruction. Additional width may also be required to ensure that appliance doors can open fully without being obstructed by adjacent appliances or units.

# 8 Using the kitchen

**Oven**

The oven should be built into a tall housing unit and set at an appropriate height for the oven door to open safely above a wheelchair user's knees without touching them. There should be sufficient work surface to the side of the oven to safely transfer a hot tray or grill pan. To assist oven use, a robust, heat resistant pull out shelf should be provided beneath the oven.

The oven should be provided with either a side hung door or a pull down door that slides away into the base of the oven. The latter is the preferred option as it allows for safer and easier use of the oven. If a side hung door is provided, it should be hinged to open away from the adjacent worktop, so dishes can be transferred directly onto the worktop without the oven door creating an obstruction.

Ovens with a pull down door that does not slide away beneath the oven should be avoided as the door acts as a barrier between a wheelchair user and the oven. A wheelchair user risks sustaining burns while reaching over the door to lift food out with extended arms.

Freestanding ovens and ovens built into units under the worktop should also be avoided, as they increase the risk of accidents and burns when reaching and lifting from a low level.

Internal oven shelves should be on telescopic runners, to prevent shelves tipping or falling out when extended.

Oven controls should allow for ease of use with reduced hand function or dexterity.

**Hob**

A hob that is fitted flush into the worktop to avoid or reduce lips or upstands provides the most suitable solution. This enables a wheelchair user to slide pans across the worktop and hob, which can reduce the risk of accidents.

Controls should be positioned to the front of the hob to allow ease of access and unhindered sliding of pans to the side.

Gas hobs should not be provided as they can increase the risk of accidents, spills or burns due to their live flame, burners and raised metal pan supports.

Induction hobs can provide the following benefits:

- safer to use as the hob remains cooler and reduces the risks of burns
- temperature adjustment is responsive
- if no pan is detected the hob will switch off
- low profile edges enable easier transferring of pans
- easier to clean.

**Extractor hood**

A wheelchair user's upper body and face will be in closer proximity to the hob than those of a standing user. The provision of an over hob extractor hood should therefore be considered as it pulls heat and steam away from the wheelchair user and can reduce the risk of burns. The extractor controls should also be accessible by the wheelchair user. Where switches cannot be located in an easily accessible position, a remotely operated extractor fan should be provided.

# Using the kitchen  8

### Fridge/freezer

Some shelves of a tall fridge/freezer are likely to be outside the reach range of a wheelchair user; however, they can still be used by other members of the household. Where there is enough space within the kitchen, separate integrated fridge and freezer appliances, set at suitable raised heights, can be considered. This provision should not reduce the overall worktop space available.

A separate fridge and freezer built under the worktop may not provide an accessible solution as they can be too low to allow a wheelchair user to reach in and lift out items.

### Dishwasher

A dishwasher can save time and effort and reduce water spillage when compared to washing up manually. Integrated drawer dishwashers can provide a more accessible solution as the appliance can be set at an optimal height for a wheelchair user.

### Washing machine and tumble dryer

Wheelchair users may have additional washing and drying requirements and need to be able to dry items quickly. A separate tumble dryer, in addition to a washing machine, is therefore beneficial.

Where possible, the washing machine and tumble dryer should be in a suitable location elsewhere in the dwelling. If located in a cupboard, they should be easily accessible and this space should be in addition to any general storage requirements.

### Switches, sockets and controls

All isolators, switches, sockets, stopcocks, radiator controls and other controls should be accessible to a wheelchair user. Refer to Chapter 12: Services and controls for further details.

### Lighting

Where possible, natural light should be maximised within kitchen layouts. Combine general lighting with well-positioned task lighting in the following areas:

- over the hob to provide a safe cooking environment
- under wall units, to reduce shadowing and provide a well-lit worktop
- within deep drawers and cupboard units.

### Tiles or wall covering

Tiles or wall covering should be easy to clean and maintain, with low reflectance.

Tiles or wall covering behind the height adjustable worktop should accommodate the full range of height adjustment and not impede movement of the worktop.

### Flooring

The floor finish in the kitchen should be slip resistant to help reduce the risk of accidents. Guidance is available from the Health and Safety Executive (www.hse.gov.uk).

Wheelchair Housing Design Guide

# 8 Technical provisions

## 8.2 Technical provisions

*When designing to Part M of the Building Regulations, Approved Document M, Volume 1, M4(3) Category 3, the technical provisions detailed in this chapter are relevant to M4(3)(2)(b) – **Wheelchair accessible dwellings**. For M4(3)(2)(a) – **Wheelchair adaptable dwellings**, the provisions within this chapter should be read in conjunction with Chapter 14: Designing wheelchair adaptable dwellings, and the provisions outlined in paragraph 14.2.5.*

**AD M**

To demonstrate that a dwelling is capable of meeting the functional and spatial provisions for a wheelchair accessible dwelling, furnished plan layouts that show the necessary clear access zones and other provisions should be shown at a scale of at least 1:100.

3.20

> **Additional good practice recommendations**
> - Provide fully furnished plans at a scale of 1:50.

### 8.2.1 Location

The kitchen and principal eating and living areas should be located on the entrance level with step-free access between these areas. The kitchen and dining area should be within the same room or connected to each other.

3.32
3.31a
3.30a
Perf – b

### 8.2.2 Basic spatial requirements

A minimum length of kitchen worktop is required in accordance with Table E, dependent on the number of bedspaces in the dwelling. This worktop length is measured through the mid-line of the worktop, not the front or rear worktop edge. Worktop length includes the sink, hob and built-in oven with the pull out shelf beneath the oven enclosure.

3.34a
Diag 3.8

Note: Full height appliances or storage units should not be included in worktop length as they reduce the worktop available for food preparation and cooking.

### Figure 8.8 Measuring worktop length

Diag 3.8

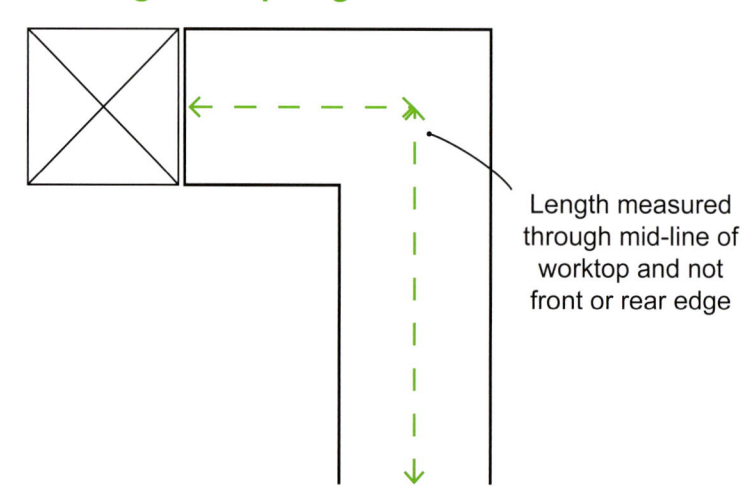

Length measured through mid-line of worktop and not front or rear edge

# Technical provisions

| Table E | Minimum length of kitchen worktop, including fittings and appliances, to be fitted at completion for a wheelchair accessible dwelling |||||
|---|---|---|---|---|---|
| Number of bedspaces* | 2 | 3 & 4 | 5 | 6–8 |
| Minimum worktop length (mm) | 6130 | 6530 | 7430 | 8530 |

AD M

Table 3.4

*For the purposes of establishing the number of bedspaces relevant to these requirements, a bedroom at or above 8.5m² and below 12.5m² is counted as one bedspace, and equal to or greater than 12.5m² as two bedspaces.*

3.36 Note 3

A minimum 1500mm wide clear access zone is required in front of, and between, all kitchen units and appliances. Any radiators, boxing-in or other localised obstructions should not project into the clear access zone.

3.32b

### 8.2.3 Worktop

The hob and combined sink and drainer unit should be provided on a continuous height adjustable section of worktop, or on a fixed section of worktop that is capable of being re-fixed at alternative heights. This section of worktop should be a minimum of 2200mm long and have clear and continuous open leg space underneath it, capable of achieving a minimum of 700mm clearance above floor level. There should be no fixed white goods (appliances) beneath this section of worktop.

3.34b

3.34b

3.34b

The height adjustable section of worktop can either be a straight run or a right angle fitted into a corner. Where the latter is provided, there should be a minimum of 300mm length of worktop between the hob and the return corner of worktop, and between the combined sink and drainer unit and the return corner of worktop.

Diag 3.8
Diag 3.8

### Figure 8.9   Height adjustable worktop in a corner

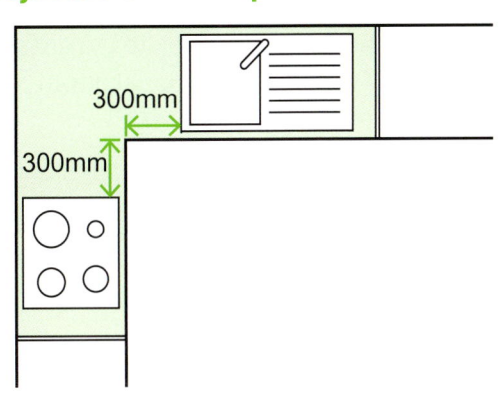

Diag 3.8

Provide a minimum 400mm of worktop to at least one side of the oven and fridge or fridge/freezer (where taller than the worktop height), or to one side of a pair of tall appliances where they are located together at the end of a run.

3.34g

Wheelchair Housing Design Guide

65

# 8 Technical provisions

**Additional good practice recommendations**

- The oven should be on the same run as the hob and combined sink and drainer unit, whether straight, L-shaped or U-shaped, without interruption by any doorways or open spaces.

- The height adjustable section of worktop incorporating the sink and drainer unit and hob should be a minimum of 2500mm to accommodate:

  – a minimum of 400mm between the edge of the hob and the edge of the height adjustable section of worktop to reduce the risk of accidents and improve usability. Where insufficient space is provided, a pan handle can catch on the adjacent fixed worktop, as the adjustable worktop moves up or down, causing spills and potential injury

  – additional space in between the hob and the sink to provide an accessible preparation area on the height adjustable section.

Figure 8.10   Height adjustable worktop

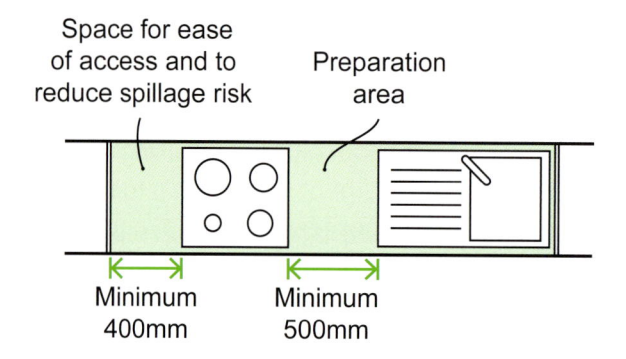

- Position all pipework towards the rear of the sink to maximise clear leg space under the height adjustable section of worktop. Additional depth of worktop may be needed to accommodate services.

- The clear leg space should not be used for storage, bins or moveable furniture.

- Provide a minimum of 600mm to at least one side of the oven and fridge or fridge/freezer, where these are taller than worktop height, to allow food or a hot dish to be easily and safely placed on the worktop.

# Technical provisions

## 8.2.4 Sink

Provide a combined sink and drainer unit, with the bowl no more than 150mm deep. The sink should have insulation to the underside to prevent scalding of a wheelchair user's legs. Taps should be lever operated and capable of easy operation.

Water supply should include isolation valves and flexible tails. Drainage should either be flexible or, where fixed, easily adaptable to allow for the worktop height to be adjusted between 700mm and 950mm above finished floor level.

AD M: 3.34c, 3.34d, 3.34h, 3.34i

> **Additional good practice recommendations**
> - Taps should be provided with a swan neck swivel arm extending over the drainer or worktop to allow a kettle or saucepan to be filled on the drainer or worktop.

## 8.2.5 Storage

> **Additional good practice recommendations**
> - Provide suitable storage within a wheelchair user's optimal reach range of 300mm to 1500mm above the floor level. Storage may incorporate tall units, base units, drawers, wall units set at a lower height, rise and fall wall units or storage with accessible pull out solutions. Integral storage space should be provided for accessible refuse and recycling.
> - Wall units fitted at 350mm above standard worktop height can allow access to the lower shelves for a wheelchair user while still enabling a microwave or other appliances to sit on the worktop.
> - Tall wall units (fitted at 350mm above standard worktop height), while not fully accessible, can increase the overall storage capacity of the kitchen.
> - Provide drawers in base units, rather than a hinged base unit door, as these are easier to access.
> - Handles should be easy to grip, have no sharp edges and be positioned to maximise ease of use.

# 8 Technical provisions

### 8.2.6 Appliances

Provide a suitable space for a built-in oven to be installed, with its centre line 1000mm above floor level.[4] A pull out shelf should be provided beneath the oven enclosure.

AD M

3.34e
3.34f

> **Additional good practice recommendations**
>
> - Provide a hob which is fitted to reduce lips or upstands, with controls positioned to the front of the worktop.
>
> - The pull out shelf beneath the oven should be heat resistant to enable hot dishes from the oven to be placed directly onto it.
>
> - The oven door should either be a pull down door that slides away into the base of the oven or a side hung door. Where a side hung door is provided, it should be hinged to open away from the adjacent worktop, so dishes can be transferred directly onto the worktop without the oven door creating an obstruction.
>
> **Figure 8.11   Oven**
>
> (a) Oven with centre line at 1000mm above floor level
> (b) Side hung oven door opening away from the counter
> (c) Telescopic oven shelf
> (d) Heat resistant pull out shelf
> (e) Minimum 600mm adjacent worktop

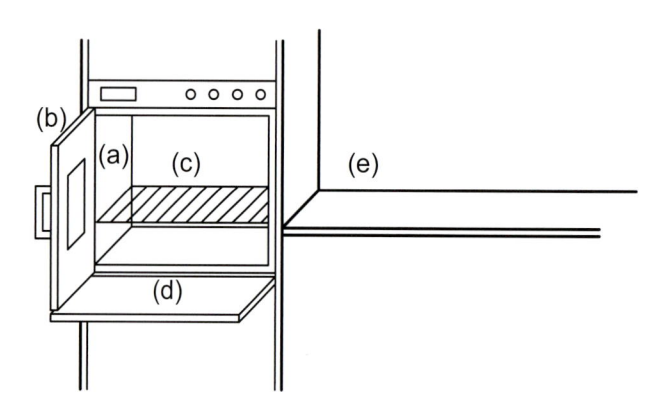

### 8.2.7 Controls and lighting

All electrical controls, including sockets, should meet the requirements of Chapter 12: Services and controls.

---

[4] Approved Document M, Volume 1 (2015 edition incorporating 2016 amendments) indicates the centre line of the oven between 800mm and 900mm above floor level. Note: The dimension given in this design guide contradicts Approved Document M; a centre line of 800mm to 900mm above floor level is not considered to be a safe oven height, as the oven door will open into a wheelchair user's legs.

3.34e

# Technical provisions

## 8.2.8 Kitchen layouts

**Figure 8.12 Example layout: three bedroom, five person dwelling**

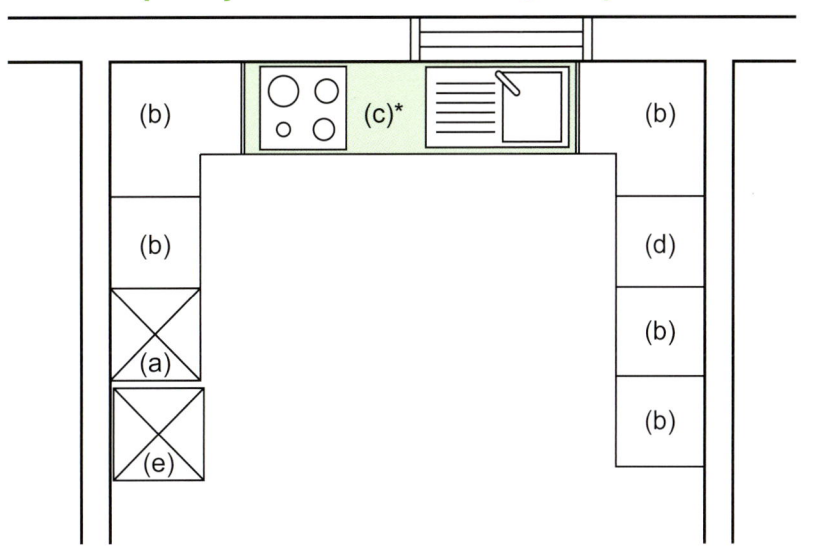

Diag 3.8

(a) Tall unit with oven
(b) Base unit/drawers/refuse/recycling
(c) Height adjustable worktop including hob and sink and drainer*
(d) Washing machine or other storage
(e) Tall fridge freezer

*Refer to additional good practice recommendations overleaf

**Figure 8.13 Example layout: one bedroom, two person dwelling**

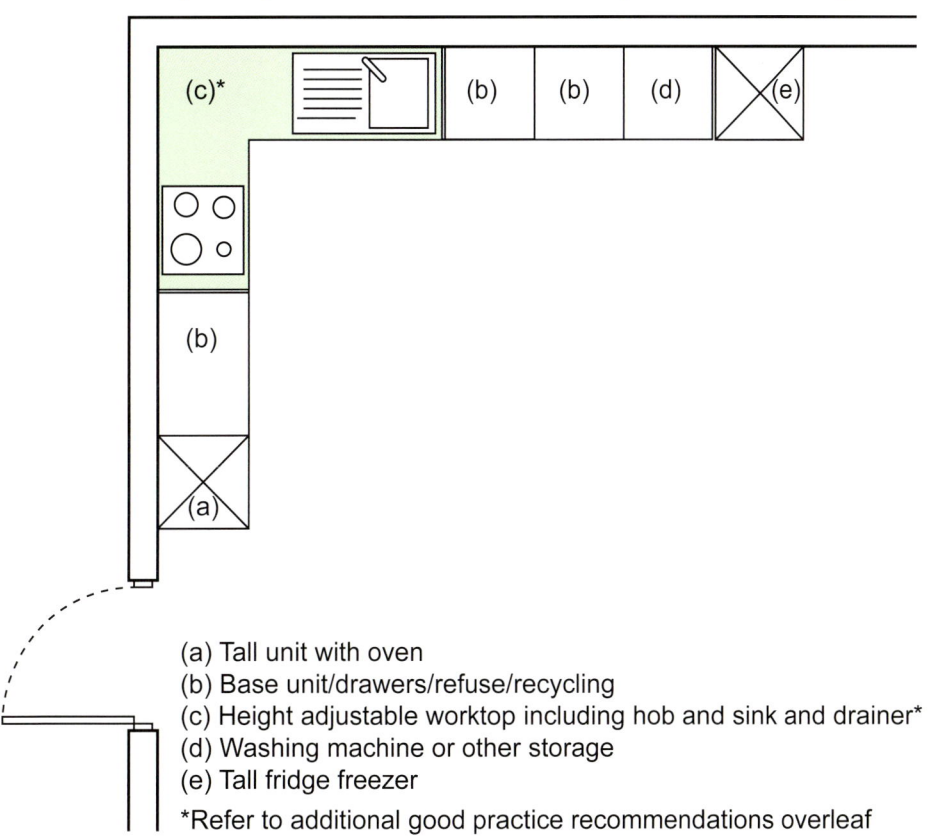

Diag 3.8

(a) Tall unit with oven
(b) Base unit/drawers/refuse/recycling
(c) Height adjustable worktop including hob and sink and drainer*
(d) Washing machine or other storage
(e) Tall fridge freezer

*Refer to additional good practice recommendations overleaf

Wheelchair Housing Design Guide

# 8 Technical provisions

**Additional good practice recommendations**

- Figure 8.12 and 8.13 show example kitchen layouts in line with AD M. To ensure optimal design for a kitchen layout, however, it is recommended that the additional good practice recommendations highlighted in this chapter are included and considered at an early design stage.

- Consider the kitchen size, shape and layout to allow for a height adjustable worktop that exceeds the minimum AD M length requirement and provide a height adjustable worktop with a minimum length of 2500mm. This height adjustable worktop should provide:

    - a minimum of 400mm between the edge of the hob and the end of the height adjustable section of worktop to improve usability and reduce the risk of accidents. Where insufficient space is provided, a pan handle can catch on the adjacent fixed worktop, as the adjustable worktop moves up or down, causing spills and potential injury

    - a minimum of 500mm between the hob and sink to provide an accessible preparation area.

### Figure 8.14   Height adjustable worktop

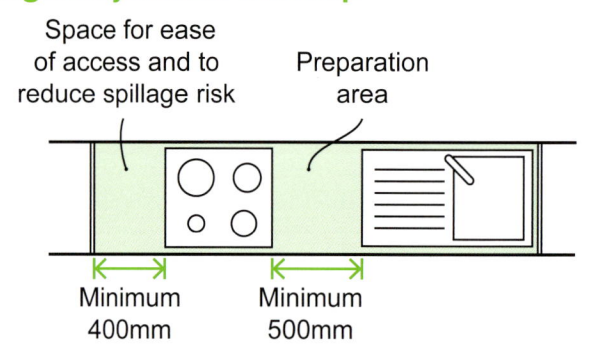

- Where possible, design the layout of the kitchen so the oven is adjacent to the height adjustable section of worktop. The clear leg space under the height adjustable worktop will then enable a wheelchair user to approach the oven at an oblique angle, making it easier to reach into the oven. This layout can only be considered, however, where the length of the height adjustable worktop exceeds the minimum AD M requirement and is provided with a minimum of 400mm between the hob and the end of the height adjustable section of worktop to allow for placement of a hot tray or grill pan from the oven.

    Note: If positioning of the oven adjacent to the height adjustable worktop is only achievable with the oven in a corner in close proximity to the return worktop/hob, then an oven with a pull down door that slides away into the base of the oven should be installed. An oven with a side hung door should not be used in this layout, as there will be insufficient space between the hot oven door and the return worktop/hob, placing the wheelchair user at increased risk of burns.

- Ensure adequate storage is provided in the kitchen, incorporating accessible storage solutions and wall units. Refer to 8.2.5 for further details.

- Minimum worktop length should be exceeded where practicable to accommodate additional storage and white goods.

# Bedrooms 9

## Principle

There should be sufficient space in all bedrooms to enable a wheelchair user to manoeuvre, transfer to the bed, reach the window and use all the furniture and controls.

## 9.1 Design considerations

In addition to the principal double bedroom, all other bedrooms should be designed for wheelchair access and use in order to provide flexibility in bedroom choice or to accommodate families with more than one disabled member. It is also important that a wheelchair user is able to access all rooms in the dwelling to complete household tasks, care for others or look after children.

Wheelchair accessible dwellings should have, as a minimum, one double bedroom. Dwellings smaller than this are often not suitable as the bedroom, associated living areas and storage space are not large enough to meet the spatial requirements of a wheelchair user. A wheelchair user living on their own may still need a double bed for ease of transfer and assistance. In some cases, an additional single bedroom may be required for personal assistants or the storage of specialist equipment.

The accessible bathroom and principal double bedroom should be located close to each other on the same floor. A wheelchair user may dress and undress in the bedroom and transfer to and from the bathroom using a shower chair, covered by a dressing gown or towel. The route between these two rooms should therefore be short and avoid passing the entrance door or living areas to offer the wheelchair user dignity and privacy.

All bedrooms should not only be sufficient in size, but should allow for the functional use of typical bedroom furniture and easy manoeuvring within the room. Locating the door away from the corner of the room, or providing doors that open beyond 90° can assist a wheelchair user to negotiate the door swing.

In each bedroom, a wheelchair user should be able to:

- enter and manoeuvre clear of the door swing and furniture
- approach and use all furniture, with sufficient space to open wardrobes and drawers
- approach both sides of a double bed or one side of a single bed, to transfer or care for children
- access all electrical controls
- approach and control windows and doors.

# 9 Bedrooms

Not all wheelchair users are able to independently transfer from their wheelchair to a bed or chair and may require support from a personal assistant or the use of specialist hoisting equipment. Where the wheelchair user is a child, transfer to the floor may also be beneficial for play or physical exercise. A portable or manual hoist is often not a suitable solution as it can be difficult to move, require a lot of manoeuvring space and possibly operation by two assistants. In many cases where hoisting is required, a tracking hoist system will provide an appropriate solution and consideration should be given to the ceiling structure from the outset to allow for ease of future adaptation.

The principal double bedroom is the room most likely to be occupied by a wheelchair user, and is typically larger than other double bedroom provisions, with additional space on each side of the bed. This allows more room for transfers, hoisting, turning of a wheelchair or care by personal assistants.

Transfer into and out of bed can take increased time and effort and some wheelchair users will require assistance. Once in bed, a wheelchair user should be able to turn the bedroom lights on and off independently and have easy access to environmental controls and switches for charging electronic devices. Access to an entry phone with a door release function adjacent to the bedhead is important to enable a wheelchair user to let in visitors or personal assistants from their bed.

The positioning of wardrobes should enable a convenient approach sideways or at a slight angle. Sliding wardrobe doors can reduce the need for a wheelchair user to negotiate a door swing and can be easier to access. Recessed built-in wardrobes may help to maximise circulation space. If providing hanging space, an adjustable rail positioned at low level is beneficial for ease of access from a wheelchair. A wheelchair user will also need sufficient space in front of drawers to approach, reverse and pull out the drawers.

Consideration should to be given at early design stages to the position of radiators to ensure they do not encroach on clear access zones or affect manoeuvrability of a wheelchair user.

# Technical provisions

## 9.2 Technical provisions

*When designing to Part M of the Building Regulations, Approved Document M, Volume 1, M4(3) Category 3, the technical provisions detailed in this chapter are relevant to both M4(3)(2)(a) – **Wheelchair adaptable dwellings** and M4(3)(2)(b) – **Wheelchair accessible dwellings**.*

To demonstrate that a dwelling is capable of meeting the functional and spatial provisions for a wheelchair user dwelling, furnished plan layouts that show the necessary clear access zones, other provisions and the furniture of the furniture schedule included in Appendix 1, should be shown at a scale of at least 1:100. — 3.20

### 9.2.1 Provision

As a minimum, each dwelling should have a principal double bedroom. — 3.35d

This principal double bedroom should either be located on the entrance storey, or the storey above or below the entrance storey, and on the same storey as the accessible bathroom described in 10.2.5 or 10.2.6. — 3.35d, 3.41b

All other bedrooms within the dwelling should also be designed to be accessible to a wheelchair user (refer to 9.2.2–9.2.6). — 3.35

> **Additional good practice recommendations**
> - There should be a short, discreet route between the principal bedroom and the accessible bathroom that does not cross through any kitchen, living or dining room or other bedroom.

### 9.2.2 Room sizes

The principal double bedroom should have a minimum floor area of 13.5m$^2$ and be a minimum of 3m wide, clear of obstructions such as radiators. — 3.35d

All other double and twin bedrooms should have a minimum floor area of 12.5m$^2$ and be a minimum of 3m wide. Any single bedroom should have a minimum floor area of 8.5m$^2$ and be a minimum of 2.4m wide. — 3.35f, 3.35i

Note: Depending on the shape of the bedroom and location of the doorway, rooms are likely to be larger than the minimum required floor area in order to accommodate furniture provision and clear access zones as described in 9.2.3 and 9.2.4.

### 9.2.3 Furniture provision

Fully furnished layouts should be shown for each bedroom, with furniture and sizes in accordance with Table F. — 3.20, 3.35 Note 1

# 9 Technical provisions

| Table F | Minimum bedroom furniture requirements |
|---|---|

**Furniture requirements applicable to the principal double bedroom, other double and twin bedrooms**

| Furniture to be shown | Furniture size (mm) | Quantity |
|---|---|---|
| Principal bedroom double bed (minimum provision) | 2000 × 1500 | 1 |
| Other double bedroom double bed; or | 1900 × 1350 | 1; or |
| Twin beds (2 single beds) | 1900 × 900 | 2 |
| Bedside table | 400 × 400 | 2 |
| Double wardrobe | 600 × 1200 | 1 |
| Chest of drawers | 450 × 750 | 1 |
| Desk | 500 × 1050 | 1 |

**Furniture requirements applicable to each single bedroom**

| Furniture to be shown | Furniture size (mm) | Quantity |
|---|---|---|
| Single bed | 1900 × 900 | 1 |
| Bedside table | 400 × 400 | 1 |
| Double wardrobe | 600 × 1200 | 1 |
| Chest of drawers | 450 × 750 | 1 |
| Desk | 500 × 1050 | 1 |

Figure 9.1  Example of principal double bedroom minimum furniture provision

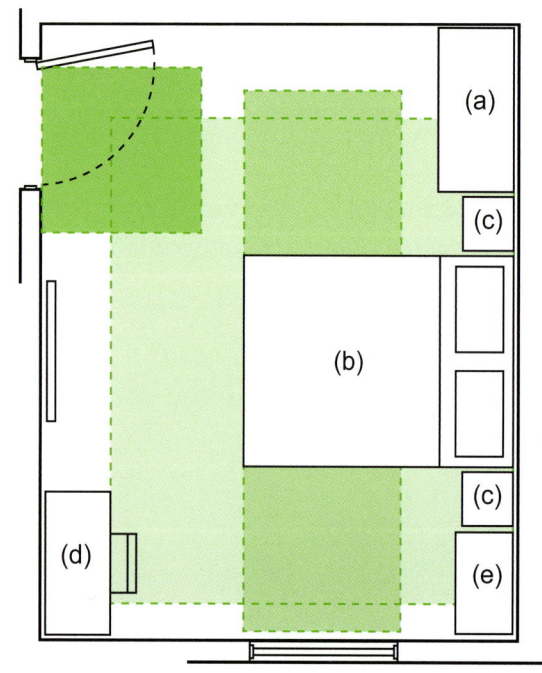

(a) Double wardrobe: 600mm × 1200mm
(b) Double bed: 2000mm × 1500mm
(c) Bedside table: 400mm × 400mm
(d) Desk: 500mm × 1050mm
    (chair permitted to overlap into access zone)
(e) Chest of drawers: 450mm × 750mm

Wheelchair Housing Design Guide

# Technical provisions

## 9.2.4 Clear access zones

All bedrooms should provide a minimum 1200mm × 1200mm manoeuvring space inside the doorway, clear of all furniture and the door (when the door is in a closed position). A minimum 750mm wide clear access route should be available from the doorway to the window.

The principal double bedroom should have a clear access zone, a minimum of 1000mm wide, to both sides and the foot of the bed. In addition, there should also be two 1200mm × 1200mm manoeuvring spaces, one to each side of the bed (refer to Figure 9.3).

Every other double bedroom should have a 1000mm wide clear access zone to one side and the foot of the bed.

In all twin and single bedrooms, a clear access zone, minimum 1000mm wide, should be provided to one side of each bed.

In all bedrooms, allow a minimum clear access zone of 1000mm in front of all furniture required by Table F. Furniture may encroach into the clear access zones by up to 600mm at the bedhead end only, and should be clear of all other access zones. Any chair provided with the desk may encroach into the clear access zones.

**AD M**
3.35b
3.35a
3.35e
3.35g
3.35h
3.35e
3.35g
3.35h
Diag 3.9
3.35e
3.35g
3.35h

### Figure 9.2 Access to bedroom furniture

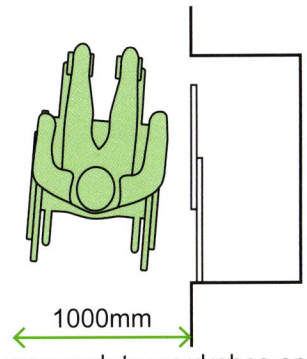

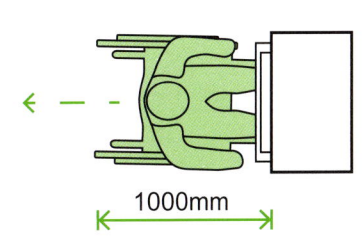

1000mm — Sideways approach to wardrobes and storage

1000mm — Space to approach, reverse and pull out drawers

Wheelchair Housing Design Guide

# 9 Technical provisions

A clear access zone can overlap with another clear access zone, but should be clear of all radiators. Refer to Table G for the minimum required bedroom access zones.

**AD M**

Diag 3.9

| Table G | Minimum bedroom access zones | |
|---|---|---|
| **Clear access zone** | **Requirement** | |
| 1 × 1200mm × 1200mm manoeuvring space inside the doorway, when the door is in a closed position | All bedrooms | 3.35b |
| 750mm wide clear access route from the doorway to the window | All bedrooms | 3.35a |
| 1000mm clear access zone in front of all furniture | All bedrooms | 3.35e<br>3.35g<br>3.35h |
| 1000mm wide clear access zone to each side of the bed and the foot of the bed | Principal double bedroom | 3.35e |
| 2 × additional 1200mm × 1200mm manoeuvring spaces; 1 to each side of the bed | Principal double bedroom | 3.35e |
| 1000mm wide clear access zone to one side and the foot of the bed | Other double bedrooms | 3.35g |
| 1000mm wide clear access zone to one side of each bed | All single and twin bedrooms | 3.35h |

### Figure 9.3  Example of principal double bedroom clear access zones

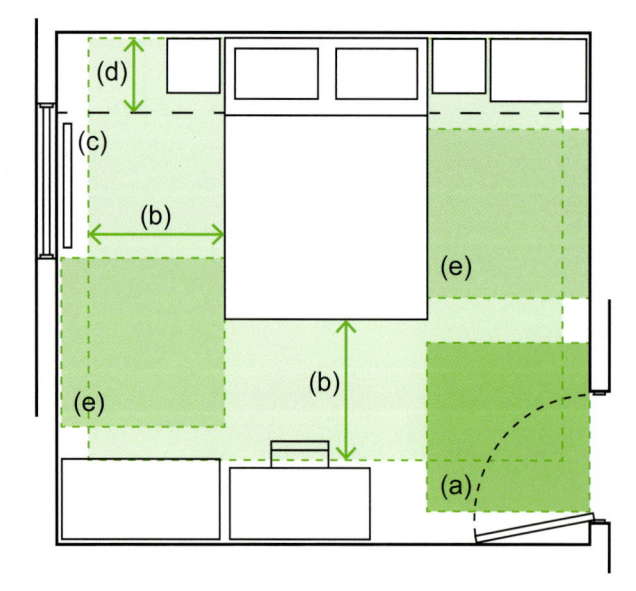

(a) Minimum 1200mm x 1200mm manoeuvring space inside the doorway, clear of the bed and door (with the door in the closed position)  — 3.35b

(b) Minimum 1000mm wide clear access zone to both sides and the foot of the bed — 3.35e

(c) Radiator placement clear of all access zones — Diag 3.9

(d) Furniture encroachment into clear access zones by up to 600mm at bedhead only — Diag 3.9

(e) Minimum 1200mm x 1200mm manoeuvring space on both sides of the bed — 3.35e

Note:
- Minimum 750mm wide access route from the door to the window — 3.35a
- Minimum 1000mm clear access zone in front of all furniture — 3.35e

Wheelchair Housing Design Guide

# Technical provisions

## Figure 9.4 Example of single room clear access zones

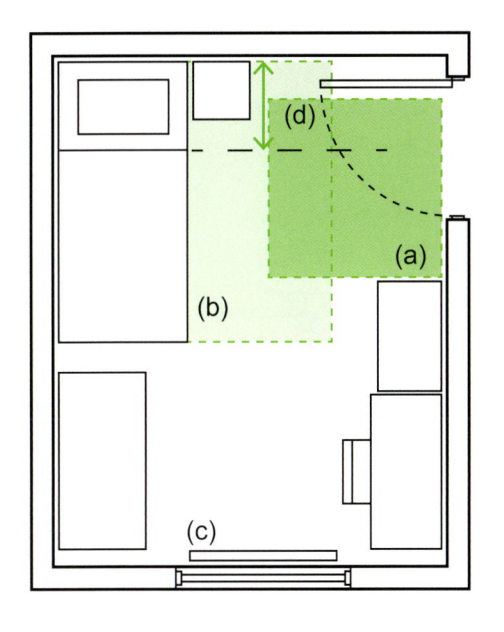

(a) Minimum 1200mm x 1200mm manoeuvring space inside the doorway, clear of the bed and door (with the door in the closed position)
(b) Minimum 1000mm wide clear access zone to one side of the bed
(c) Radiator placement clear of all access zones
(d) Furniture encroachment into clear access zones by up to 600mm at bedhead only

Note:
- Minimum 750mm wide access route from the door to window
- Minimum 1000mm clear access zone in front of all furniture

AD M
3.35b
3.35h
Diag 3.9
Diag 3.9
3.35a
3.35h

### Additional good practice recommendations

- Provide a minimum 1500mm turning circle instead of the 1200mm × 1200mm manoeuvring space inside the doorway (with the door in a closed position) to allow greater manoeuvrability on entering/leaving the room and negotiating the door swing. This is likely to impact minimum bedroom sizes and bedroom layouts should be considered in the early design stages to ensure that clear access zones and furniture requirements can be accommodated.

- Provide 1350mm in front of the chest of drawers to allow sufficient space to approach, reverse and pull out drawers.

- Provide a 1000mm wide access route, instead of a 750mm wide access route, from the bedroom door to the window and to any external doors from the bedroom.

- Profiling or height adjustable beds or beds with pressure relief mattresses vary in size and can be as long as 2300mm. Allow for this where possible, ensuring a clear 1000mm at the foot of the bed.

- Provide 1500mm × 1500mm manoeuvring space to one side of the bed, to enable enough space for a personal assistant(s) to assist with a hoisted transfer. This space should be provided towards the upper half of the bed.

### 9.2.5 Controls

In the principal bedroom, make provision for the future installation of controls grouped adjacent to the bedhead. This could be achieved by providing blank sockets, conduit and draw wires. The bedhead controls should include:

- TV aerial and power socket outlets
- two-way light switch
- telephone and broadband sockets.

3.44j

Wheelchair Housing Design Guide

# 9 Technical provisions

| | AD M |
|---|---|
| A door entry phone with remote door release should be provided in the principal bedroom, adjacent to the bedhead. | 3.44i |
| Switches and controls should be located with the centre line of any switch or control button(s) 700mm to 1000mm above floor level and a minimum of 700mm (measured horizontally) from any inside corner. Refer to Chapter 12: Services and controls. | 3.44b |

### Figure 9.5  Bedhead controls

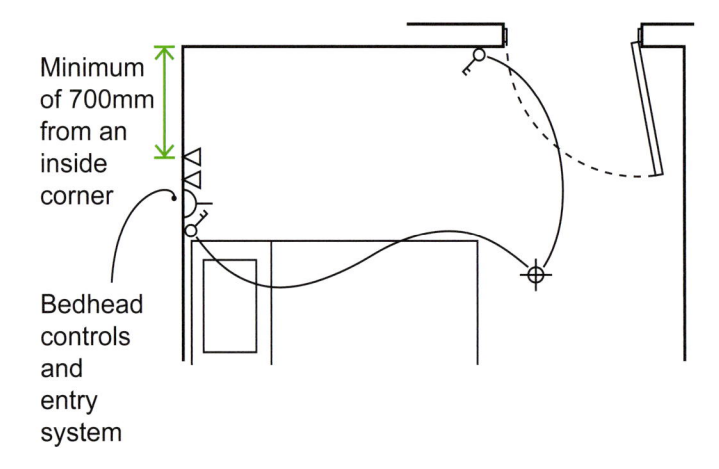

3.44b
3.44i
3.44j

*Minimum of 700mm from an inside corner*

*Bedhead controls and entry system*

---

**Additional good practice recommendations**

- In the principal bedroom, provide the controls grouped adjacent to the bedhead (refer to the list above) from the outset.
- Include sockets with USB charging points.

---

#### 9.2.6  Future hoist provision

The ceiling structure to all bedrooms should be strong enough to allow for the fitting of an overhead hoist capable of carrying a load of 200kg. (Additional localised strengthening may still be required to support high point loads at the time that adaptations are fitted.)

3.35c
3.35 Note 2

# Bathrooms 10

## Principle

The ability to access and use toileting and washing facilities is essential and the design of the bathroom should enable independence and dignity for wheelchair users. Space and layout should also take into account the potential need for support from a personal assistant, as well as the needs of the household.

## 10.1 Design considerations

The number of WC and bathroom facilities should suit the size of the household and take account of the additional time a wheelchair user may need to use the facilities. In larger dwellings, more than one WC will be of benefit to accommodate the needs of a wheelchair user, their family and any visitors.

In dwellings of more than one storey, a wheelchair user should have access to a WC/cloakroom on the entrance level and an accessible bathroom on the same floor level as their bedroom.

In all dwellings, an accessible bathroom should be provided on the same floor and close to the principal bedroom. This is important for dignity, privacy and ease of use, as a wheelchair user may dress and undress in the bedroom and transfer to and from the bathroom using a shower chair, covered by a dressing gown or towel. Where there is only one accessible bathroom, sole access should not be via a bedroom, e.g. an en-suite layout. This allows more flexibility as it may not be known at the design stage who the disabled family member(s) is.

An accessible bathroom layout should provide sufficient space for a wheelchair user and a personal assistant to use all the facilities. There should be sufficient circulation space to easily enter the bathroom, manoeuvre and turn around. Clear access zones should be provided to each bathroom fitting, with sufficient clear space for a wheelchair user to approach the fitting and transfer to the WC, shower and/or bath.

Most wheelchair users are likely to find it easier to access and use a level access shower rather than a bath. Wheelchair accessible dwellings should therefore provide an installed level access shower from the outset, although there should still be sufficient space to install a bath in place of the shower if required.

In larger dwellings, both a level access shower and an accessible bath should be provided. A wheelchair user may take longer to use the facilities and this enables another family member to bath or shower at the same time. For convenience and privacy, the bath and shower should ideally be provided in separate rooms.

In the early design stages, it is critical to identify horizontal and vertical plumbing and drainage, including vertical stacks and horizontal ducts on drawings. These should be boxed in to avoid exposed pipework and should not impact on circulation space or

# 10 Bathrooms

encroach into required access zones. Bathrooms may need to be larger than anticipated to accommodate pipework and drainage.

To minimise the risk of scalding, all water supply features should be provided with appropriate temperature control.

Outward opening bathroom doors increase space for manoeuvring inside the room and can assist access/egress in an emergency, such as a person falling behind the door. A rail or handle on the back of the door can assist a wheelchair user to close the door behind them and should be fitted according to individual need. Refer to Chapter 11: Operating internal doors and windows.

### WC

Both sides of the WC should be clear of all obstructions, such as stacks, drainage and pipework. The WC should project sufficiently to enable a commode/shower chair to be wheeled back far enough over the WC to align with the WC aperture and to assist with lateral transfers.

Sufficient space to the side, diagonally and in front of the WC is also important to accommodate a range of transfers. The wash hand basin should therefore not be positioned directly alongside or in front of the WC where it will obstruct this transfer zone.

### Figure 10.1    Range of transfers to a WC

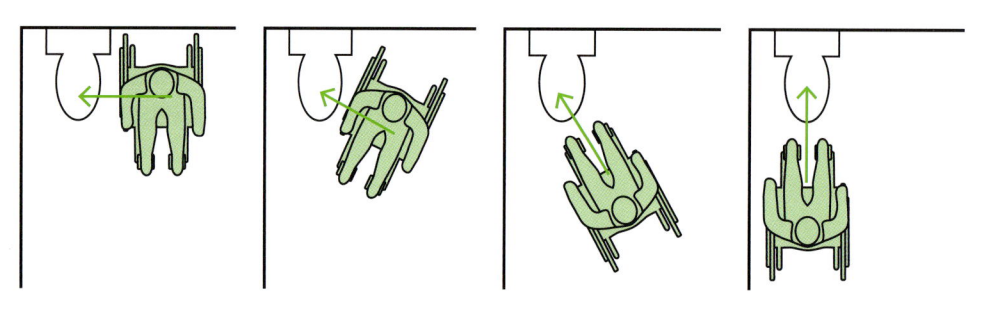

Where there is a fall in the floor to assist with drainage of the level access shower, this should not form part of the transfer zone to the WC, as an uneven transfer zone could present a tipping hazard.

WC flush handles should be large and easy to use, e.g. a lever flush handle. They should be positioned on the front of the cistern on the outer, or transfer side, so a wheelchair user can reach the flush once they have transferred off the WC back to their wheelchair. Where a sensor flush is fitted, it should be positioned for similar ease of access and reach.

Consideration should be given to the provision of an isolated electrical supply (in line with relevant Building Regulation) for the future installation of an automatic toilet, such as those that wash and dry.

### Shower

Where a level access shower is installed, the bathroom should be constructed as a wet room. A corner shower design is the most suitable layout as it provides easier and safer access for an independent wheelchair user, or someone providing assistance. It is

# Bathrooms 10

important that the shower and its controls to be positioned on the adjacent wall to the potential shower seat position for easy reach. If the controls, shower riser bar and wall-mounted shower seat are positioned along the same wall, it is difficult for the person sitting to turn and reach behind them. This type of layout also makes it difficult to effectively enclose and contain water within the showering area.

### Figure 10.2  Shower fittings

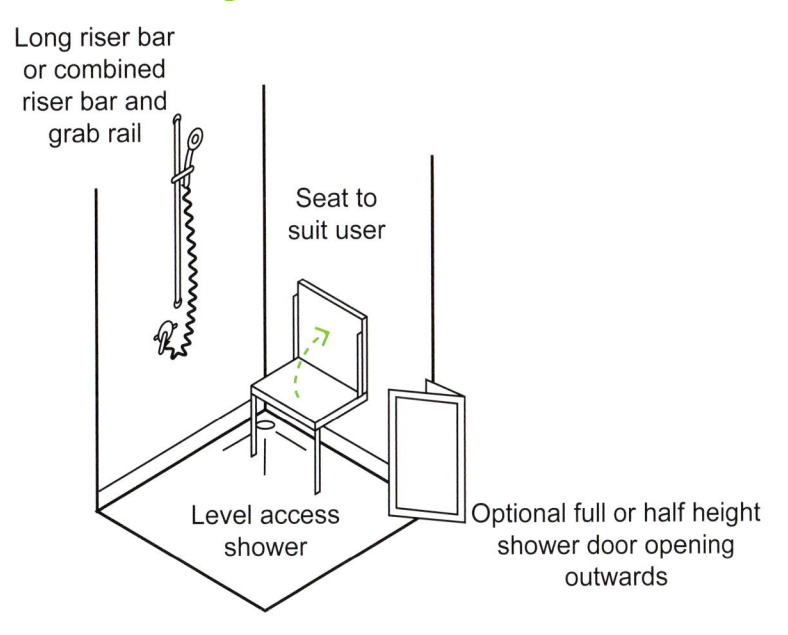

*Long riser bar or combined riser bar and grab rail*

*Seat to suit user*

*Level access shower*

*Optional full or half height shower door opening outwards*

Level access showers should be step-free with no lips or upstands and a floor laid with shallow falls towards a floor gulley/grate connected to the drainage. Pre-formed shower trays recessed into the floor (rather than graded screed) ensure a uniform gradient and reduce the potential for pooling of water. Recessed trays should be covered in the same slip resistant flooring as the rest of the bathroom.

Containment of water is important to keep the area outside the shower as dry as possible. To avoid water egress, the shower and its drainage should be situated in the corner furthest away from the door. Full length, weighted shower curtains can assist with water containment and prevent clinging to the user but should be of good quality and machine washable. If half or full height folding shower doors are fitted, they should not restrict access to the shower or movement within the bathroom and should be easy to operate and fold back against the wall. Fixed shower screens should be avoided as they are likely to obstruct access and circulation space and can restrict access for a personal assistant.

Shower controls should be easy to access, understand and operate. Control of water temperature and volume of flow should be possible with one hand. Consideration should be given to how a seated person will reach the controls, as well as how a person assisting would operate the shower without getting wet. Showers which can be operated remotely prevent an independent wheelchair user from having to sit under cold water when the shower is first turned on.

Wheelchair Housing Design Guide

# 10 Bathrooms

To enable a person seated, standing or assisting to reach and use a hand held shower, an extended length shower riser bar should be fitted in an appropriate position to accommodate all users. A combined shower riser bar and grab rail should be considered as they reduce the need for additional support rails, can appear less institutional and reduce the likelihood of a person pulling on the riser bar. There should be provision for a range of shower head positions which are easy to adjust. Where a fixed wall or ceiling mounted shower head is fitted, an additional hand held shower and riser bar should be provided. The location of a fixed shower head should also take account of the wheelchair user's position while taking a shower.

Shower hoses should be anti-syphon and of sufficient length for a person assisting to reach and use.

**Bath**

A wheelchair user may require specialist equipment to access the bath, or support them while in the bath. Baths that are non-sculptured with flat rims will accommodate a wider range of bathing equipment. The depth of the bath should also be considered, as shallow baths or those with low overflow outlets can restrict the practical use of equipment. The construction and materials used may make some baths unsuitable to accommodate the extra weight of bathing equipment.

Taps should be easy to reach and operate, such as lever type. They should not be mounted on the wall side of the bath, as these can be difficult to reach. Recessed integral hand grips may also be of benefit but should not restrict the fitting of bathing equipment. If an over-bath shower is provided, it should be accessible to someone seated or standing in the bath.

Where there is boxing-in at the end of the bath, it should be load bearing to accommodate a seated person, have no sharp edges and be sealed to prevent water ingress. This boxing-in may be convenient to use as a seat when dressing/undressing; however, it should not be seen as a means to transfer in and out of the bath as a person is unlikely to be able to get back out of the bath safely. There should be sufficient clear circulation space alongside the full length of the bath for a wheelchair user to transfer.

**Basin**

The wash hand basin should be wall hung, with sufficient clear space in front of the basin for a wheelchair user to approach. Some adjustability in height for future adaptation is desirable so flexible plumbing should be considered. Where traps are unprotected they should be insulated to prevent scalding. Semi-pedestal basins, with sufficient wheelchair access underneath, can be considered to provide protection from pipework and can appear less institutional.

Wash hand basins should provide a reasonably sized bowl with space around the bowl to place toiletries. Basins with wide, flat rims and rounded edges can enable a wheelchair user to carry out a range of activities, place items safely and rest their arms more comfortably. Taps may be separate or a mixer fitting but should be easily manipulated (e.g. lever taps). Plugs should also be easy to use and a pop-up waste requiring access to the rear of the taps should be avoided where possible.

A tall mirror should be fitted above the basin so that a person seated or standing can see themselves in it. A short section of splash back tiles can be provided between the rim of the

# Bathrooms

basin and the bottom of the mirror. Consider fitting demisting heat pads to the rear of mirrors, as a wheelchair user may not be able to reach the mirror to wipe it.

If lights, switches or shaver sockets are provided alongside the mirror or basin, these should be in a wheelchair accessible location. A longer light pull cord should be provided to basin lights where necessary.

### Supports

Bathroom fittings, including wall fixings, should be specified for quality and robustness, as basins and WCs in particular may be leant on by a wheelchair user for additional support.

The walls, ducts and boxing to all WC and bathroom facilities should be suitably constructed to allow grab rails, shower seats and any other adaptations to be safely and securely fitted. The ceiling structure to these facilities should also be strong enough to allow for the fitting of ceiling track hoisting if required.

### Heating

Effective heating is important as wheelchair users may take longer in the bathroom. It can also assist with drying the floor and therefore potentially reduce the risk of slips and falls.

Radiators or heated towel rails should be carefully positioned outside clear access zones, so they do not compromise circulation space and are less likely to be used for support. Low surface temperature (LST) radiators should be provided, or radiators should be protected, to reduce the risk of scalding or burns. LST covers can be bulky or rust in wet room settings and alternative heat sources should be considered, such as an LST towel rail, wall-mounted fan heater with a long pull cord or underfloor heating. Where underfloor heating is fitted, it should take account of the shower position. It is beneficial for the bathroom heating to be controlled independently from the rest of the system.

### Floor finishes

The floor finish in bathrooms should be slip resistant. Slip resistant flooring may reduce the risk of accidents and improve the stability of wheeled equipment. Guidance is available from the Health and Safety Executive (www.hse.gov.uk).

### Wall finishes

Wall tiling or panelling should extend well beyond all wet areas. Consideration should be given to providing tonal contrast against fixtures and fittings while also reducing reflection and glare.

### Accessories

Provision should be made for all necessary bathroom accessories, such as toilet roll holders, towel rails, mirrors, hooks, cabinets and shelving to accommodate toiletries. These should be fitted within easy reach, ensuring that their positioning does not affect the safe use of bathroom fittings or encroach on clear access zones.

# 10 Technical provisions

## 10.2 Technical provisions

*When designing to Part M of the Building Regulations, Approved Document M, Volume 1, M4(3) Category 3, the technical provisions detailed in this chapter are relevant to M4(3)(2)(b) – **Wheelchair accessible dwellings**. For M4(3)(2)(a) – **Wheelchair adaptable dwellings**, the provisions within this chapter should be read in conjunction with Chapter 14: Designing wheelchair adaptable dwellings, and the provisions outlined in paragraph 14.2.6.*

To demonstrate that a dwelling is capable of meeting the functional and spatial provisions for a wheelchair accessible dwelling, plan layouts that show the necessary clear access zones, other provisions and the sanitary fittings of the furniture schedule included in Appendix 1, should be shown at a scale of at least 1:100.

| AD M |
| --- |
| 3.20 |

> **Additional good practice recommendations**
> - Provide plans at a scale of 1:50.
> - Plans are required for Building Regulations approval; however, it is recommended that fully furnished layouts are designed at planning stage to ensure that an accessible and functional layout can be achieved.

### 10.2.1 Overview

Dwellings should provide suitable WC and washing facilities. The minimum requirements for sanitary facilities vary according to the number of bedspaces and storeys within a dwelling.

3.36
Table 3.5

Each sanitary fitting (WC, basin, bath and shower) has its own associated clear access zone that should be accommodated within the room layout. These clear access zones can overlap each other unless noted otherwise.

Diag 3.11
Diag 3.13
3.43a

Every dwelling requires a bathroom with an installed level access shower as described in 10.2.5 or 10.2.6 on the same storey as the principal bedroom. Depending on the number of bedspaces, there should either be sufficient space to fit a bath in place of the installed level access shower, should this be required in future (refer to 10.2.5), or to install a useable bath in addition to the installed level access shower (refer to 10.2.6).

3.41b
3.41a
3.43c

Where a dwelling contains more than one storey, and the bathroom with the installed level access shower described in 10.2.5 or 10.2.6 is not located at entry level, the entrance storey requires a WC/cloakroom constructed as a wet room with a WC, basin and installed level access shower meeting the requirements of 10.2.8.

3.36c
3.37a

Dwellings with four or more bedspaces require access to a minimum of two WCs in separate bathrooms or WC/cloakrooms (refer to Tables H and I).

3.36b

An outward opening door is required for all wheelchair accessible sanitary facilities.

3.37c
Diag 3.12
Diag 3.15
Diag 3.16
Diag 3.17

# Technical provisions 10

### 10.2.2 Minimum requirement for sanitary provision in typical dwelling types[5]

Provide sanitary facilities in accordance with Table H and Table I, based on the number of bedspaces and storeys within a dwelling, e.g. in a four bedspace, two storey dwelling, provide as a minimum an entrance storey WC/cloakroom with installed level access shower and a bathroom with installed level access shower on the same storey as the principal bedroom.

Note: Table H and Table I provide a summary of requirements and should be read in conjunction with the detailed requirements in paragraphs 10.2.3 to 10.2.8.

| Table H | Summary of minimum requirements for sanitary provision in single storey dwellings based on number of bedspaces | | | |
|---|---|---|---|---|
| **Single storey dwellings** (typically a flat or bungalow) | | **Number of bedspaces*** | | |
| | | **2 & 3** | **4** | **5 or more** |
| **Description of sanitary facility** | | **Number of required facilities** | | |
| **Bathroom with installed level access shower:** A wet room with WC, basin and an installed level access shower. Incorporate sufficient space for a potential bath, with the relevant clear access zone (if not provided elsewhere in the dwelling), to be fitted in place of the shower (refer to 10.2.5) | | 1 | 1 | |
| **Bathroom(s) with installed level access shower and useable bath:** A wet room with WC, basin and both an installed level access shower and useable bath fitted from the outset. These facilities can be provided in one bathroom or in more than one bathroom (refer to 10.2.6) | | | | 1 |
| **Second WC/cloakroom provision:** As a minimum, a WC/cloakroom with a WC and basin (refer to 10.2.7) | | | 1 | 1 |

*\* For the purposes of establishing the number of bedspaces relevant to these requirements, a bedroom at or above 8.5m² and below 12.5m² is counted as one bedspace, and equal to or greater than 12.5m² as two bedspaces.*

**AD M**

Table 3.5

Table 3.5

3.37a
3.41a
3.36c

3.37a
3.43c
3.36c

3.37b
3.40

3.36 Note 3

---

[5] Approved Document M, Volume 1 (2015 edition incorporating 2016 amendments) provides a summary of minimum requirements that may cover other dwelling configurations.

Wheelchair Housing Design Guide

# 10  Technical provisions

| Table I | Summary of minimum requirements for sanitary provision in two or three storey dwellings based on number of bedspaces | | | | AD M |
|---|---|---|---|---|---|
| | | | | | Table 3.5 |
| **Two or three storey dwellings where the principal bedroom is not on the entrance storey (typically a house or maisonette)** | | **Number of bedspaces*** | | | |
| | | **2 & 3** | **4** | **5 or more** | |
| **Description of sanitary facility** | | **Number of required facilities** | | | |
| **Entrance storey WC/cloakroom with installed level access shower:** A WC/cloakroom constructed as a wet room with a WC, basin and installed level access shower (refer to 10.2.8) | | 1 | 1 | 1 | 3.37a 3.36c |
| **Bathroom with installed level access shower:** A wet room with WC, basin and an installed level access shower, on the same storey as the principal bedroom. Incorporate sufficient space for a potential bath, with the relevant clear access zone (if not provided elsewhere in the dwelling), to be fitted in place of the shower (refer to 10.2.5) | | 1 | 1 | | 3.41a 3.41b 3.36c |
| **Bathroom(s) with installed level access shower and useable bath:** A wet room with WC, basin and both an installed level access shower and useable bath fitted from the outset. These facilities can be provided in one bathroom or in more than one bathroom on the same storey as the principal bedroom (refer to 10.2.6) | | | | 1 | 3.43c 3.41b 3.36c |

*For the purposes of establishing the number of bedspaces relevant to these requirements, a bedroom at or above 8.5m² and below 12.5m² is counted as one bedspace, and equal to or greater than 12.5m² as two bedspaces.*

3.36 Note 3

Note: The provisions within this chapter do not apply to sanitary facilities that are additional to the minimum provisions of Table H and Table I.

3.36 Note 2

### 10.2.3  Minimum size of sanitary fittings

The size of all sanitary fittings should be in accordance with Table J. Where larger sanitary fittings are provided, these should not reduce the clear access zones required in Figures 10.3.1 to 10.3.5, 10.6.1 to 10.6.3 and 10.8.1 to 10.8.4.

Appx D
Diag 3.11
3.43a

# Technical provisions

**10**

AD M

| Table J | Minimum size of sanitary fittings |||
|---|---|---|---|
| **Bathroom sanitary fittings** || **WC/cloakroom sanitary fittings** ||
| **Sanitary fitting** | **Size (mm)** | **Sanitary fitting** | **Size (mm)** |
| WC (pan and cistern) | 500 × 700 | WC (pan and cistern) | 500 × 700 |
| Wash hand basin | 600 × 450 | Hand rinse basin | 350 × 200 |
| Level access shower | 1200 × 1200 | Level access shower | 1000 × 1000 |
| Bath | 700 × 1700 | | |

Appx D
Diag 3.11

### 10.2.4  Access zones

Each sanitary fitting (WC, basin, bath and shower) has its own associated clear access zone that should be accommodated within the relevant room layout. A 1500mm diameter clear turning circle should be provided in bathroom(s) and WC/cloakrooms with a shower. This turning circle can encroach into the shower area by a maximum of 500mm.

3.39a
Diag 3.11
Diag 3.13
3.43e

An access zone can overlap with another clear access zone, but should still be clear of sanitary fittings, radiators, towel rails and services unless otherwise indicated. Stacks or soil and vent pipes should be clear of all access zones and should not be placed in the access zone adjacent to the WC pan and cistern.

Diag 3.11
Diag 3.13

Refer to Figures 10.3.1 to 10.3.5, 10.6.1 to 10.6.3 and 10.8.1 to 10.8.4.

### 10.2.5  Dwellings of up to four bedspaces: Bathroom with installed level access shower

A bathroom with an installed level access shower should be provided on the same storey as the principal double bedroom. This should be constructed as a wet room and contain a WC, basin and installed level access shower, incorporating the access zones in Figures 10.3.1 to 10.3.5 and a clear turning circle with a minimum diameter of 1500mm.

3.41b
3.36c
3.41a
3.39a
Diag 3.11
3.43e

In dwellings of up to four bedspaces, a level access shower should be provided as the default provision but there should be sufficient space to fit a bath, and its clear access zone, in place of the shower if required.

3.41a
3.43b

Wheelchair Housing Design Guide

# 10 Technical provisions

AD M
Diag 3.11
3.43a

### Figure 10.3.1  WC access zone (bathrooms)

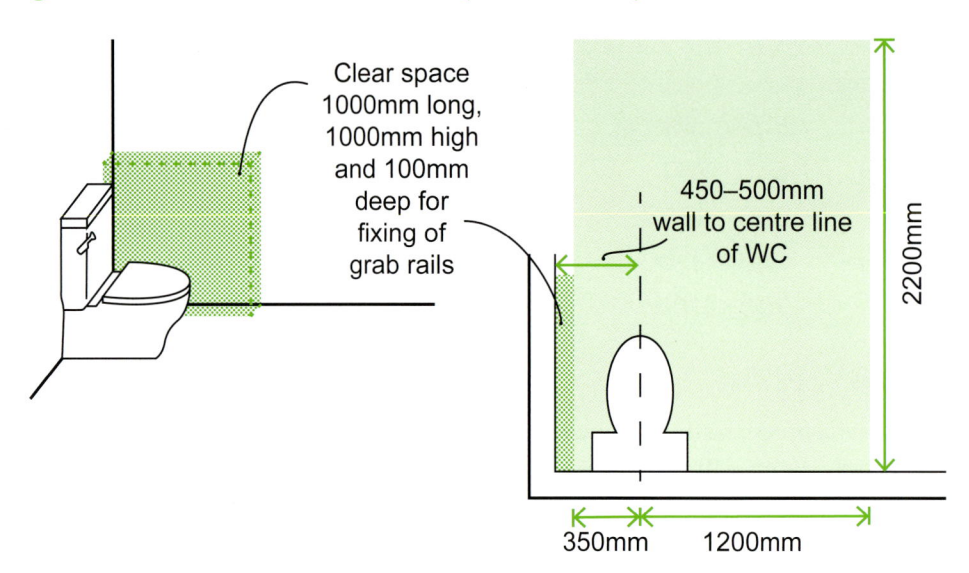

Clear space 1000mm long, 1000mm high and 100mm deep for fixing of grab rails

450–500mm wall to centre line of WC

2200mm

350mm  1200mm

### Figure 10.3.2  Permitted encroachment into WC access zone (bathrooms)

Diag 3.11
3.43a

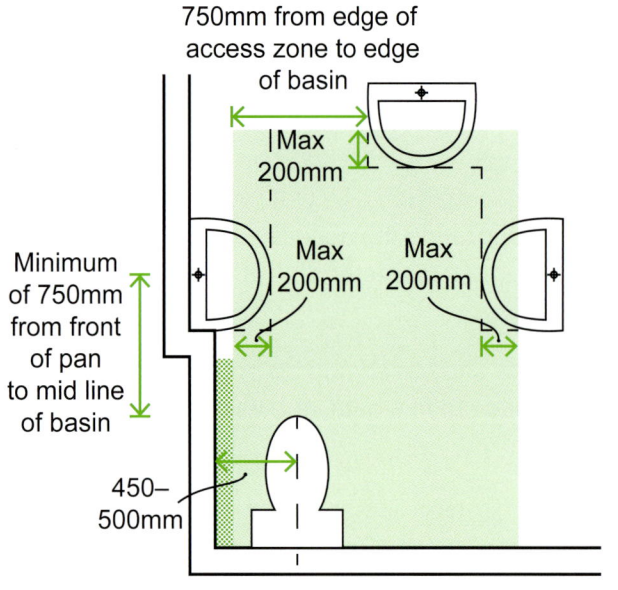

Minimum of 750mm from edge of access zone to edge of basin

Max 200mm

Minimum of 750mm from front of pan to mid line of basin

Max 200mm   Max 200mm

450–500mm

450–500mm

Wash hand basin can encroach into the WC access zone for a maximum of 200mm, with options for basin positions anywhere inside the dotted lines

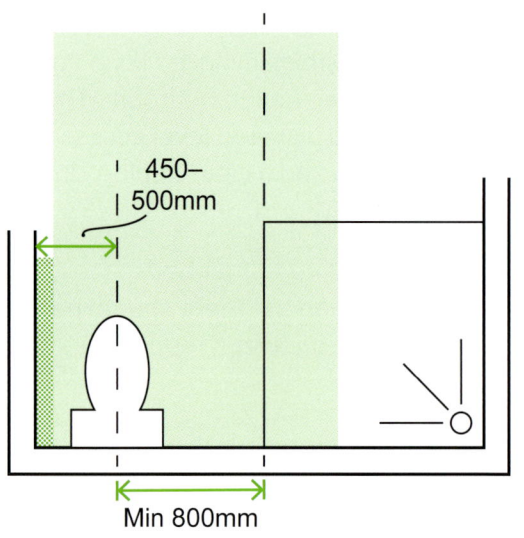

Min 800mm

The shower can encroach into the WC access zone, provided there is a minimum of 800mm between the centre line of the WC and the edge of the shower

Wheelchair Housing Design Guide

# Technical provisions

## 10

**AD M**
Diag 3.11
3.43a

**Figure 10.3.3   Wash hand basin access zone (bathrooms)**

800mm

1650mm

**Figure 10.3.4   Bath access zone (bathrooms)**

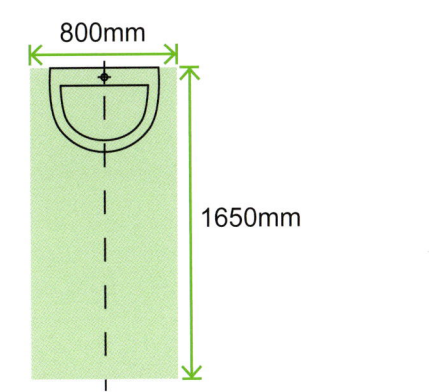

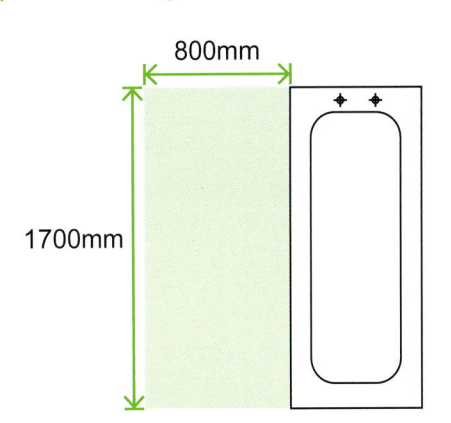

800mm

1700mm

**Figure 10.3.5   Level access shower access zone (bathrooms)**

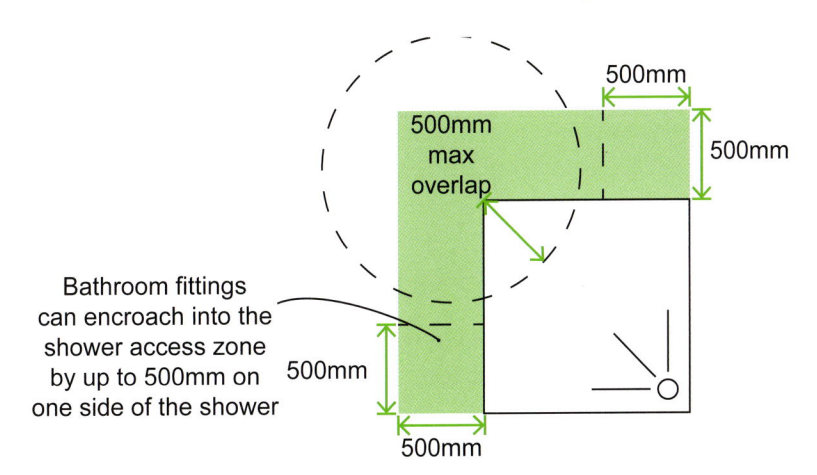

Diag 3.11
3.43a

500mm

500mm max overlap

500mm

Bathroom fittings can encroach into the shower access zone by up to 500mm on one side of the shower

500mm

500mm

Wheelchair Housing Design Guide

89

# 10 Technical provisions

### Figure 10.4  Example layout – bathroom with installed level access shower

AD M
Diag 3.11
Diag 3.16

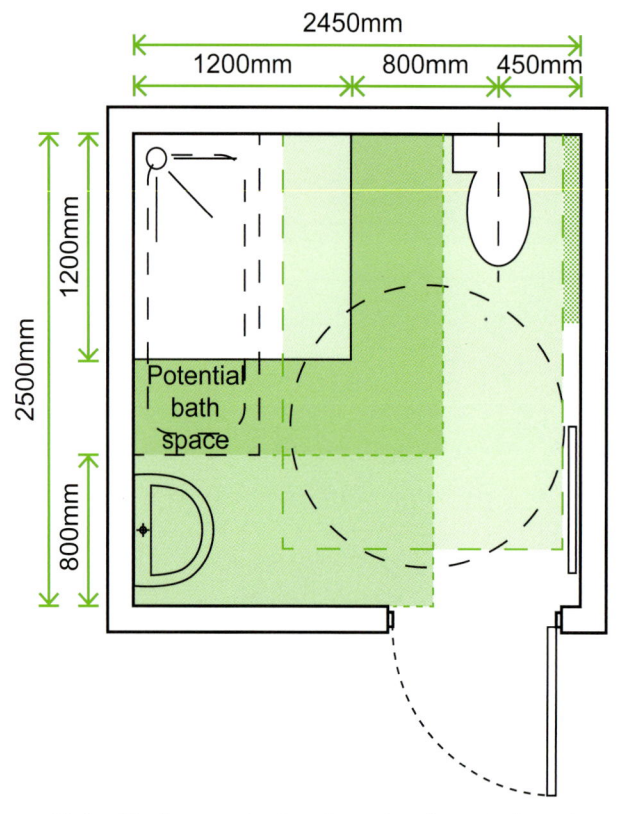

Note: Drainage, ductwork and soil vent pipes have not been shown, but will need to be accommodated outside the required access zones

**Additional good practice recommendations**

- Provide a minimum of 1200mm between the centre line of the WC and the edge of the shower. Many wheelchair users will use this space next to the WC to transfer, and a level transfer surface will reduce the risk of accidents.

### 10.2.6  Dwellings of five or more bedspaces: Bathroom(s) with installed level access shower and useable bath

In dwellings of five or more bedspaces a WC, basin and both an installed level access shower and a useable bath should be fitted from the outset, with their associated clear access zones as required in Figures 10.3.1 to 10.3.5 and a clear turning circle of 1500mm minimum diameter.

3.43c
3.39a
Diag 3.11
3.43e

The installed level access shower and useable bath can be provided in one bathroom or in more than one bathroom on the same storey as the principal bedroom. Where both a bathroom and a shower room are provided, it would be acceptable for one, but not both of these rooms, to be an en-suite bathroom. In this instance, the separate room providing the bath is permitted smaller access zones as outlined in Figure 14.1.1 to 14.1.4.

3.43c
3.41b
3.41 Note 1
3.41 Note 2

# Technical provisions

**10**

| | AD M |
|---|---|
| The room that contains the installed level access shower should be constructed as a wet room. | 3.36c |
| **Figure 10.5  Example layout – bathroom with installed level access shower and useable bath** | Diag 3.11<br>Diag 3.17 |

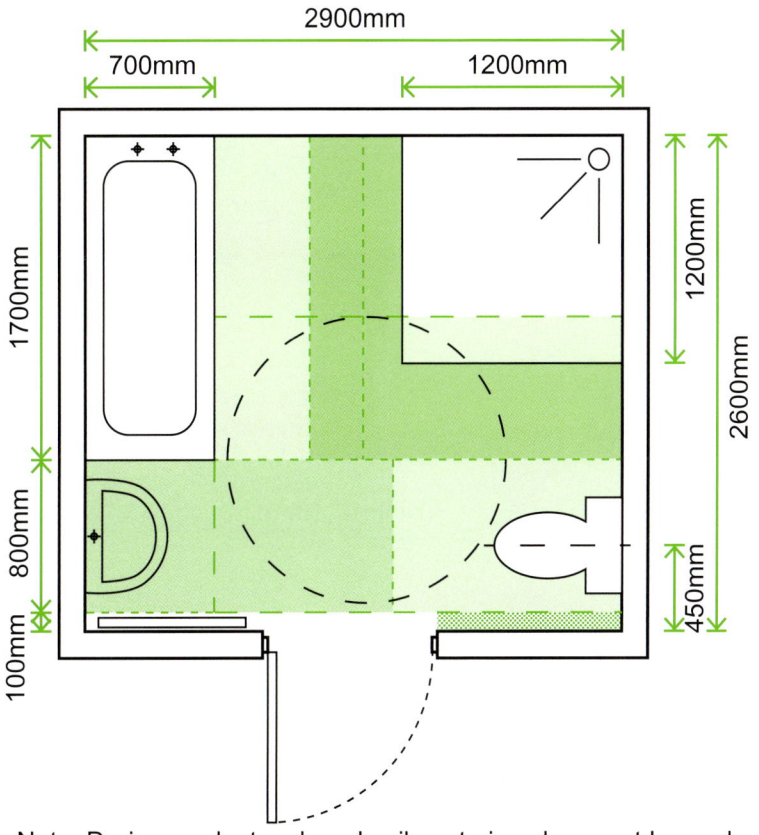

*[Figure 10.5: Plan view of bathroom layout. Overall dimensions 2900mm wide × 2600mm deep. Bath 1700mm along left wall with 700mm width dimension shown. Shower area 1200mm × 1200mm in top right. WC on right side with 450mm dimension. Basin below bath with 800mm and 100mm dimensions. Turning circle shown in centre. Door swings into room.]*

Note: Drainage, ductwork and soil vent pipes have not been shown, but will need to be accommodated outside the required access zones

### 10.2.7  Second WC/cloakroom provision (single storey dwellings)

| | |
|---|---|
| In single storey dwellings of four or more bedspaces, a second WC/cloakroom is required in addition to the bathroom with the level access shower, as described in 10.2.5 or 10.2.6. The WC, basin and their associated clear access zones of the second WC/cloakroom facility should, as a minimum, meet the requirements of Figures 10.6.1 to 10.6.3. | 3.36b<br>3.37b<br>3.40<br>Diag 3.13 |

Wheelchair Housing Design Guide

# 10 Technical provisions

**Figure 10.6.1 WC access zone (second WC provision)**

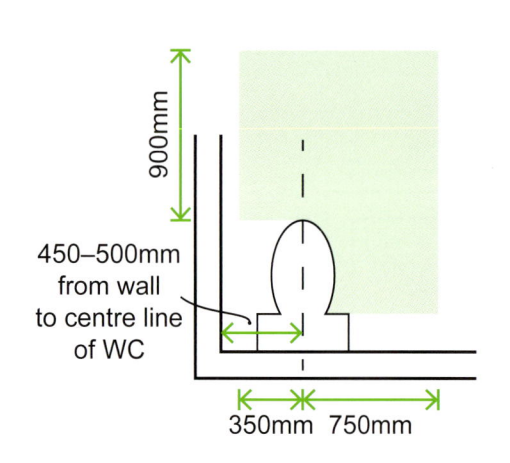

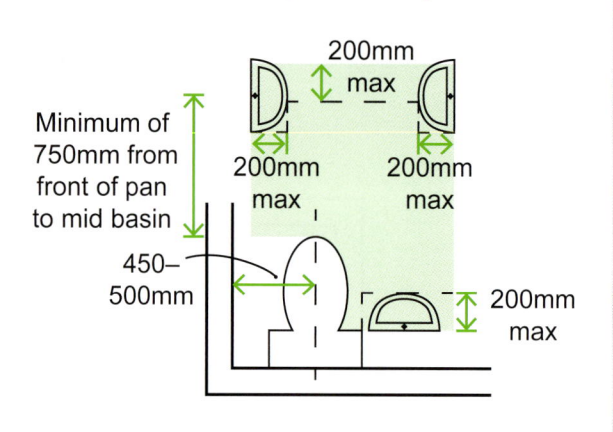

900mm
450–500mm from wall to centre line of WC
350mm 750mm

**Figure 10.6.2 Permitted hand rinse basin encroachment into WC access zone (second WC provision)**

200mm max
Minimum of 750mm from front of pan to mid basin
200mm max  200mm max
450–500mm
200mm max

AD M
Diag 3.13

**Figure 10.6.3 Hand rinse basin access zone (second WC provision)**

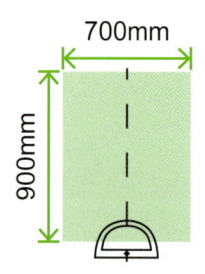

700mm
900mm

Access zone extends under the basin as far as any pedestal or vanity unit

Diag 3.13

**Figure 10.7 Example layout (second WC provision)**

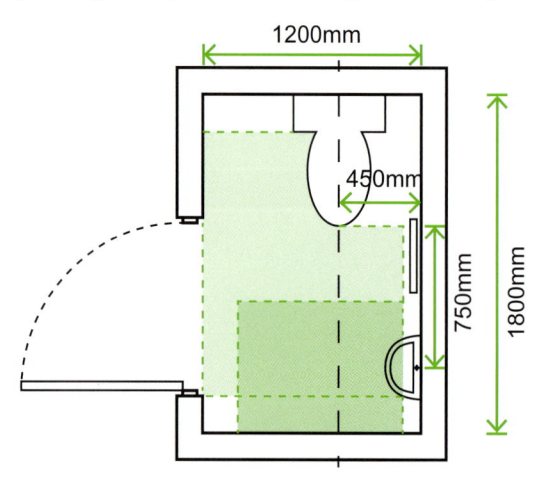

1200mm
450mm
750mm
1800mm

Note: Drainage, ductwork and soil vent pipes have not been shown, but will need to be accommodated outside the required access zones

Diag 3.13
Diag 3.14

Wheelchair Housing Design Guide

# Technical provisions

## 10.2.8 Entrance storey WC/cloakroom with installed level access shower

Where the bathroom with the installed level access shower, as described in 10.2.5 or 10.2.6, is not on the entrance storey, a WC/cloakroom with installed level access shower is required as a minimum.

This should be constructed as a wet room and contain a WC, basin and an installed level access shower, incorporating the access zones in Figures 10.8.1 to 10.8.4 and a clear turning circle of minimum diameter 1500mm. The door to this WC should open outwards.

### Figure 10.8.1  WC access zone (entrance storey WC/cloakroom)

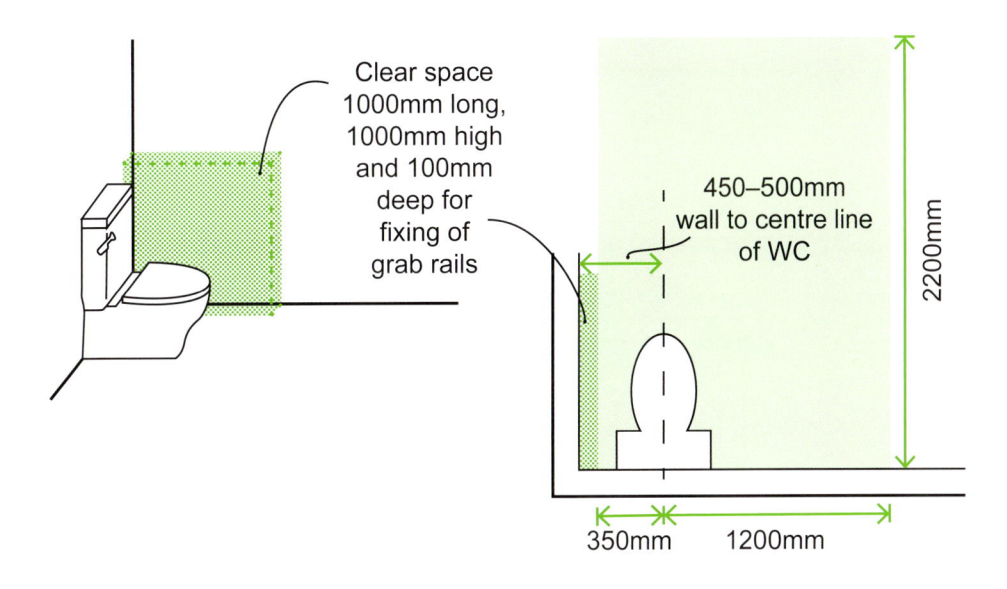

Clear space 1000mm long, 1000mm high and 100mm deep for fixing of grab rails

450–500mm wall to centre line of WC

2200mm

350mm  1200mm

### Figure 10.8.2  Permitted encroachment into WC access zone (entrance storey WC/cloakroom)

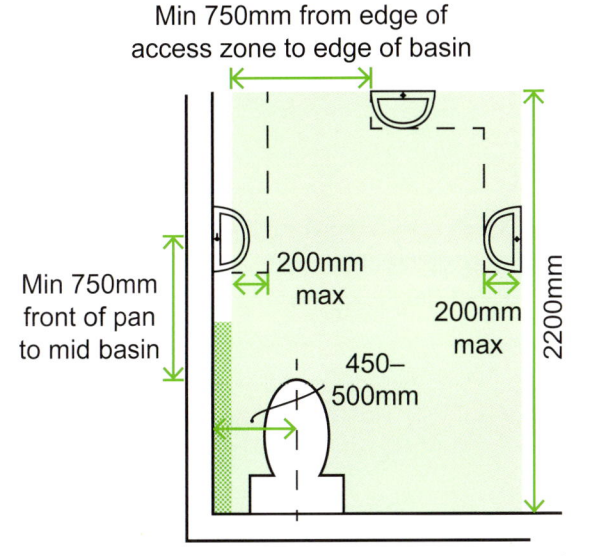

Min 750mm from edge of access zone to edge of basin

Min 750mm front of pan to mid basin

200mm max

200mm max

450–500mm

2200mm

450–500mm

1000mm

Hand rinse basin can encroach into the WC access zone for a maximum of 200mm with options for basin positions anywhere inside the dotted lines

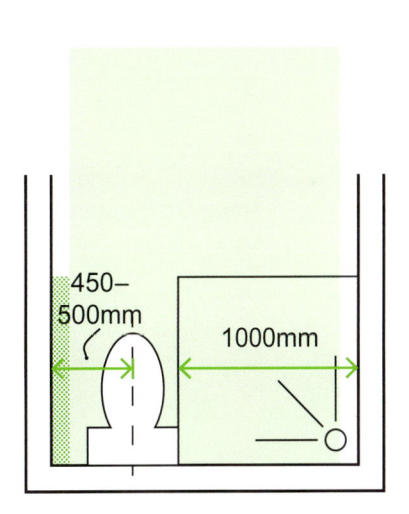

The shower can encroach into the WC access zone as shown

AD M

3.37a

3.36c
3.39a
3.37c

3.39a
Diag 3.11

3.39a
Diag 3.11

Wheelchair Housing Design Guide

# 10 Technical provisions

**Figure 10.8.3 Basin access zone (entrance storey WC/cloakroom)**

**Figure 10.8.4 Shower access zone (entrance storey WC/cloakroom)**

AD M
3.39a
Diag 3.11

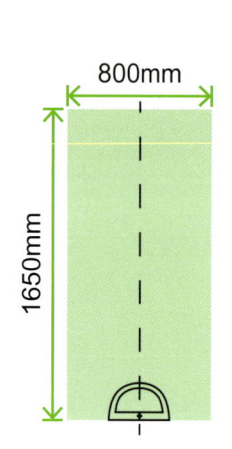

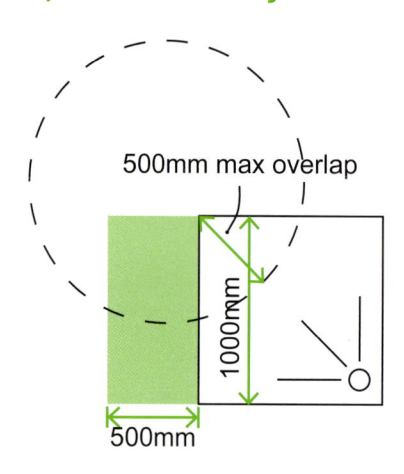

800mm
1650mm

500mm max overlap
1000mm
500mm

**Figure 10.9 Example layout – entrance storey WC/cloakroom**

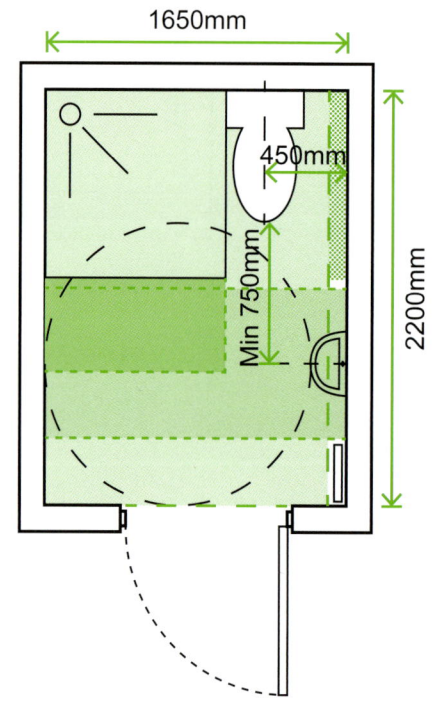

Diag 3.11
Diag 3.12

1650mm
450mm
Min 750mm
2200mm

Note: Drainage, ductwork and soil vent pipes have not been shown, but will need to be accommodated outside the required access zones

94  Wheelchair Housing Design Guide

# Technical provisions

## 10.2.9 Access from the bedroom

> **Additional good practice recommendations**
>
> - Where only one accessible bathroom is provided in a dwelling, the sole access to the bathroom should not be via a bedroom.

## 10.2.10 WC

WC pans should be a minimum of 400mm high. Whether close-coupled or wall mounted, the WC should project a minimum of 700mm from the rear wall.

3.36g
Appx D

WC flush controls should be positioned on the front of the cistern on the transfer side so they are easy to reach. Controls should be easy to grip, e.g. a lever flush handle.

3.36f

> **Additional good practice recommendations**
>
> - A WC with a pan height a minimum of 420mm from finished floor level will assist a wheelchair user to transfer onto the WC. Height should be measured from finished floor level to the top of the pan and exclude the seat.
> - Where possible, allow for 750mm projection from the front of the pan to the rear wall. This ensures that a commode/shower type chair can be wheeled far enough back over the WC to align with the WC aperture.

## 10.2.11 Shower

Installed level access showers should be step-free with no lips or upstands with a floor laid to shallow falls towards a floor gulley connected to the drainage system.

3.37a

The level access shower should be positioned in a corner to allow a wall-mounted shower seat to be fitted on the adjacent wall to the shower controls. The shower controls should be easy to operate single handed.

3.43d
3.44l

# 10 Technical provisions

> **Additional good practice recommendations**
>
> ● As a minimum, an accessible hand held shower with a 1m shower riser bar should be provided within easy reach from a seated or standing position.
>
> **Figure 10.10    Shower layout**
>
> *[Diagram showing shower layout with dimensions: 600–650mm from corner to shower riser centre, 1000mm riser bar length, centre line of seat 500mm from corner, 700–750mm, 800–900mm height to shower controls, height of seat to suit user.]*
>
> Note: Grab rails fitted to suit user

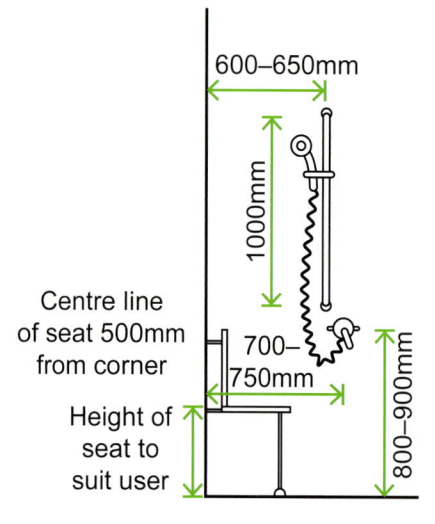

### 10.2.12  Bath

The bath provided should be a minimum of 1700mm × 700mm. Bath taps should be easy to operate single handed.

Appx D
3.44l

### 10.2.13  Basin

Basins should be wall hung, with their rim 770mm to 850mm above finished floor level. The clear zone underneath the basin should be maximised to enable wheelchair users to approach and use the basin. The access zone underneath the basin should be no lower than 600mm from finished floor level. A semi pedestal or insulated trap are permitted provided they are no lower than 400mm from finished floor level.

3.36h
Diag 3.11

Taps should be easy to operate single handed.

3.44l

# Technical provisions

**10**

### Figure 10.11   Clear access zone under the basin

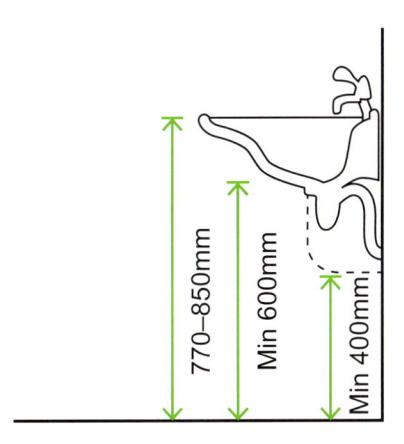

**AD M**
3.36h
Diag 3.11

**Additional good practice recommendations**

- Position the rim of the basin between 800mm and 850mm above finished floor level to allow easier access to the basin.

### 10.2.14   Services

Where located in sanitary facilities, stacks, vents, drainage, boxing-in and radiators should be clear of access zones.

Diag 3.11
Diag 3.12
Diag 3.13
Diag 3.14
Diag 3.15
Diag 3.16
Diag 3.17

**Additional good practice recommendations**

- Ensure that the position of all stacks, vents, drainage and boxing-in is shown on bathroom layouts at planning stage. Where these are located within bathrooms, the overall size of the bathroom may need to be larger to ensure that the required clear access zones (refer to 10.2.4) can be provided free from obstructions.

### 10.2.15   Finishes

**Additional good practice recommendations**

- Provide wall tiling around sanitary fittings and ensure that the tiled areas extend sufficiently beyond wet areas.
- Install slip resistant flooring in line with Health and Safety Executive guidance.

Wheelchair Housing Design Guide

# 10 Technical provisions

| | AD M |
|---|---|

**10.2.16 Supports**

The ceiling structure to sanitary facilities required by Tables H and I should be strong enough to allow the fitting of an overhead hoist capable of carrying a load of 200kg. — 3.36e

The walls, ducts and boxings to the sanitary facilities required by Tables H and I should be strong enough to support the fitting of grab rails, shower seats and other adaptations that could impose a load of up to 1.5kN/m². Additional localised strengthening may be required if adaptations are fitted that impose higher point loads. — 3.36d, 3.36 Note 1

# Operating internal doors and windows 11

## Principle

All internal doors and windows should be easy to approach, reach and operate.

## 11.1 Design considerations

**Doors**

The choice of ironmongery should be carefully considered to increase the ease with which a wheelchair user is able to operate doors within the dwelling. Ironmongery should be easily identifiable within the doorset and align with adjacent controls, such as light switches, which can improve identification and use.

All door handles and locks should be easy to grip and operate single handed. Door handles should be a lever or 'D' type, with a suitable profile and extension, and a sufficient gap between the door face and the handle. Handles which require turning, such as doorknobs, should be avoided as they are very difficult for many people to grip and turn. Door handles and locks should be spaced sufficiently far apart to allow for single handed operation.

Any sliding doors should have non-recessed pull handles for convenient operation from either side, combined with locking where needed. The handle should be easy to reach and operate when the door is in the fully open position.

Where locks are provided they should be easy to manipulate with minimal force. It should be possible to release lockable doors from both sides to allow for emergency access.

An additional rail or handle on the back of a door can assist a wheelchair user to pull the door closed behind them when moving through, e.g. on the bathroom door. The height of this rail should be set for individual use and can reduce the reach and force required to close the door. Door construction should therefore allow for retrospective fitting of grabrails.

# 11 Operating internal doors and windows

**Figure 11.1 Grab rails on back of door**

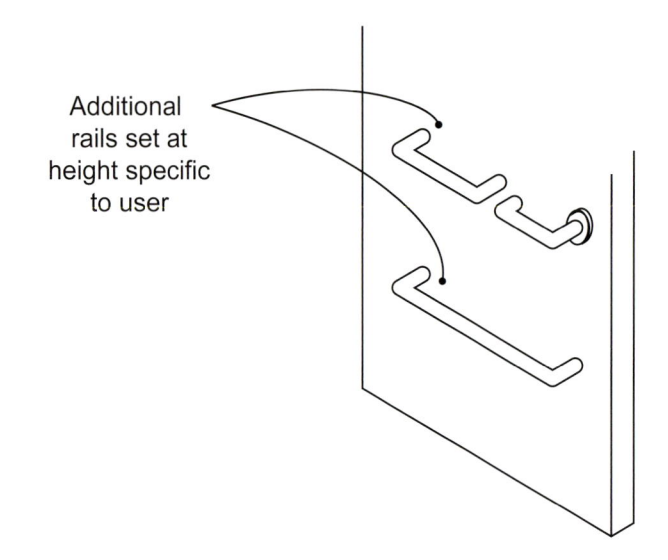

Additional rails set at height specific to user

Where door closers are necessary, they should require minimum force to open and close.

**Windows**

Wheelchair users should be able to approach, reach and operate window handles in all rooms. A sideways approach to a window enables easier reach than a head-on approach.

Where window handles cannot be placed at an accessible height, or the handle itself is not easy to operate, remote opening devices can be used. These should, however, be avoided where possible due to aesthetics and maintenance issues. Where passive ventilation devices, such as trickle vents, are required they should be duplicated at a lower level so they are accessible to a wheelchair user.

# Technical provisions

**11**

**AD M**

## 11.2 Technical provisions

*When designing to Part M of the Building Regulations, Approved Document M, Volume 1, M4(3) Category 3, the technical provisions detailed in this chapter are relevant to both M4(3)(2)(a) –* **Wheelchair adaptable dwellings** *and M4(3)(2)(b) –* **Wheelchair accessible dwellings**.

### 11.2.1 Internal doors

All doors should have handles, locks, latches and catches that are easy to grip and use, fitted 850mm to 1000mm above floor level.

3.44f

### Figure 11.2 Door handle height

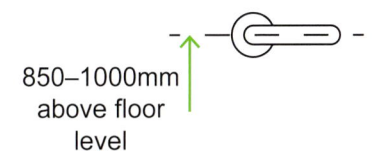

3.44f

850–1000mm above floor level

Also refer to Chapter 5: Dwelling circulation areas and storage.

> **Additional good practice recommendations**
> 
> - Bathroom doors within the dwelling should be easily unlocked from the outside in the event of an emergency.
> 
> - If door closers are used within the dwelling they should not require an opening force more than 30N from 0° to 30° or more than 22.5N from 30° to 60° of the opening cycle.

### 11.2.2 Windows

The handle to at least one window in the principal living area should be 700mm to 1000mm above floor level. Handles to all other windows should be 450mm to 1200mm above floor level. Where these handle heights are not achieved, the window should be fitted with a remote opening device within this height range.

3.44d
3.44e

Wheelchair Housing Design Guide

## 11 Technical provisions

**Figure 11.3   Window handle height in the principal living area**

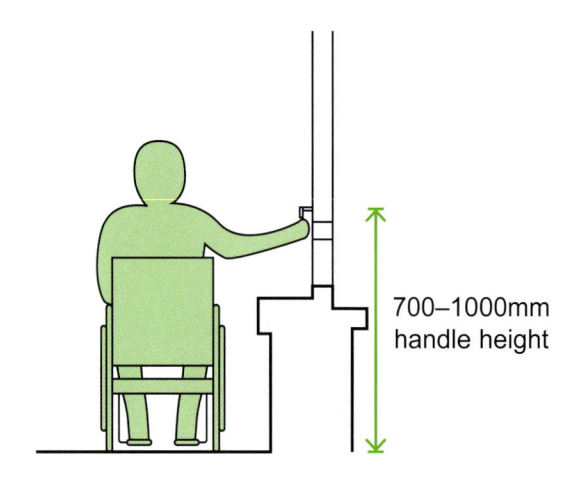

AD M
3.44d

700–1000mm handle height

Also refer to Chapter 7: Using living spaces and Chapter 9: Bedrooms.

> **Additional good practice recommendations**
> - All window handles, locks, latches and catches should be easy to grip and use single handed.

# Services and controls    12

## Principle

All services and controls should be suitably located to ensure they are reachable by all users and designed to be easily operated and understood.

## 12.1  Design considerations

Access to services and controls within the dwelling can be facilitated by careful selection of fittings and their location.

**Switches and sockets**

Switches and sockets should always be sited for convenient access at a suitable height and away from corners. It is difficult for a wheelchair user to reach into corners, as the wheelchair user's feet and footplates protrude and restrict access. They should also not be positioned behind appliances or doors, where they would be difficult to use.

**Figure 12.1    Positioning of switches and sockets**

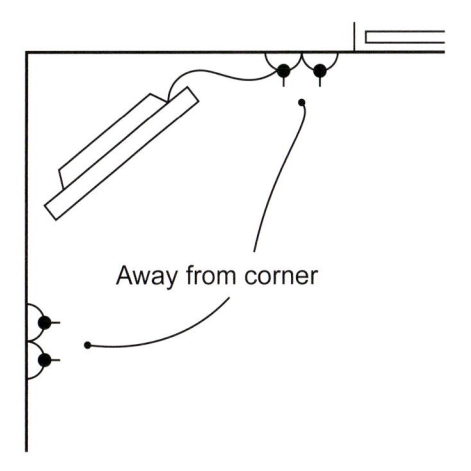

Away from corner

Where switches are small, or more than one switch is provided on a plate, these can be difficult to use. Switches should therefore be provided on a full plate, wide rocker or individual plates. Switches to double socket outlets should be located at the outer ends of the plate.

Pull cord switches to any appliances or lighting should be a suitable length, have larger pulls for ease of grip and restraining eyelets to ensure that the cord stays within reach.

Consideration should be given to providing two-way switching where appropriate. A two-way switch is useful in the bedroom, so the light can be switched on when entering the room and turned off once in bed. This is also of benefit on staircase landings and in living areas.

Wheelchair Housing Design Guide

# 12 Services and controls

Wheelchair users may require additional sockets to accommodate or charge specialist equipment, reduce the need to unplug and change socket use, or to use ever evolving home control systems. Sockets with integrated USB charging ports should be considered in bedrooms, kitchen and living areas. A good standard of distributed daylight and additional sockets for task lighting are also of benefit. These reduce reliance on ceiling or wall-mounted lights, where a wheelchair user may not be able to replace lightbulbs independently.

**Consumer units**

Consumer units should be located in a suitable position so that they do not obstruct circulation space or reduce minimum corridor widths. Where consumer units are located within a cupboard, they should be placed within easy reach.

**Stopcocks**

Stopcocks should not be positioned behind appliances or in the back of cupboards. Where they are not within easy reach of a wheelchair user, a remote water switch should be provided.

**Entry phone and security**

A door entry phone should be provided with remote door release functions in the living area and principal bedroom. This will assist wheelchair users who find it more difficult and time consuming to answer the door, or who may need to allow access for a personal assistant without getting out of bed. A video entry system and other smart technology solutions can enhance security.

**Provision in bedrooms**

Switches, sockets for equipment, TV aerials, telephone, broadband and the door entry phone should all be adjacent to the bedhead in the principal bedroom.

**Heating**

Heating should be responsive, controllable and capable of achieving an even temperature throughout the house. Some wheelchair users need higher temperatures for personal comfort and may be susceptible to a drop in temperature as they move between rooms. Heating controls should be easy to reach, operate and understand.

Where individual radiators are provided, the controls should be within the wheelchair user's reach. Low surface temperature radiators should be provided in bathrooms and cloakrooms to avoid scalds and burns when a wheelchair user is undressed or has impaired sensation. Also refer to Chapter 10: Bathrooms.

**Technology**

Environmental controls and other assistive technology allow many wheelchair users to increase their independence at home. This type of technology is continually evolving and new developments and products should be incorporated into the design of wheelchair user dwellings where practicable.

# Technical provisions

## 12.2 Technical provisions

*When designing to Part M of the Building Regulations, Approved Document M, Volume 1, M4(3) Category 3, the technical provisions detailed in this chapter are relevant to both M4(3)(2)(a) – **Wheelchair adaptable dwellings** and M4(3)(2)(b) – **Wheelchair accessible dwellings**, with the exception of the location of isolators to kitchen appliances and height requirements for radiator controls. For wheelchair adaptable dwellings also refer to Chapter 14: Designing wheelchair adaptable dwellings, and the provisions outlined in paragraph 14.2.7.*

### 12.2.1 Switches, sockets and controls

Services and controls should be positioned to assist wheelchair users who have reduced reach. | 3.44 Perf – e

Boiler timer controls and thermostats should either be mounted 900mm to 1200mm above finished floor level on the boiler, or separate controllers (wired or wireless) should be mounted elsewhere in an accessible location within the same height range. Radiator controls should be mounted 450mm to 1000mm above floor level. | 3.44m / 3.44n

All other switches, sockets, stopcocks and controls should be located with their centre line 700mm to 1000mm above floor level and a minimum of 700mm (measured horizontally) from an inside corner, and not positioned behind appliances. Kitchen appliances should have isolators located within this same height range. | 3.44b / 3.44c

Note: Pull cord switches to bathroom appliances such as shaver light fittings or extractor fans should be provided with the cord 700mm to 1000mm above floor level.

### Figure 12.2 Positioning of switches | 3.44b

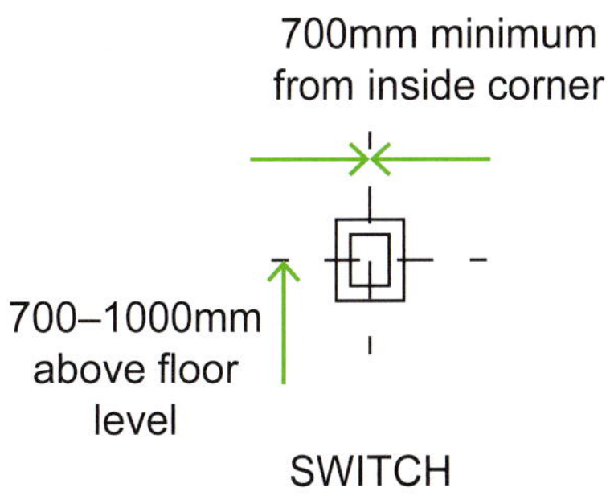

700mm minimum from inside corner

700–1000mm above floor level

SWITCH

Light switches should be on individual plates unless wide rocker or full plates are provided. | 3.44g

Switches to double socket outlets should be located at the outer ends of the plate (rather than in the centre). | 3.44h

# 12 Technical provisions

### Figure 12.3  Switches and sockets

Full plate switch

Socket outlet switched at outer ends

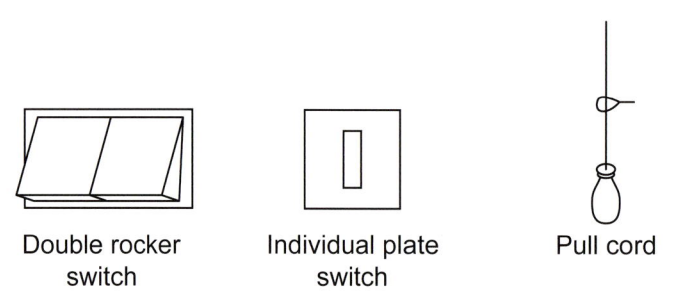

Double rocker switch

Individual plate switch

Pull cord

AD M
3.44g
3.44h

A power socket should be provided in the wheelchair transfer and storage space. A power socket suitable for powering a stair-lift should be provided close to the foot or head of a stair to which a stair-lift might be fitted. In dwellings of more than one storey, a power socket suitable for powering the lifting device should be provided close to the liftway.

3.25c
3.30d
3.29d

A telephone point and a main electrical power socket should be provided together in the main living space.

3.44k

In the principal bedroom, make suitable provision to install bedhead controls in the future, grouped adjacent to the head of the bed, e.g. by providing blank sockets, conduit and draw wires. These should include:

3.44j

- TV aerial and power socket outlets
- two-way light switch
- telephone and broadband sockets.

A door entry phone with remote door release facility should be provided in the main living space and the principal bedroom. In the principal bedroom, the door entry phone should be grouped with the provision outlined above for future bedhead controls.

3.44i
3.44j

### Figure 12.4  Bedhead controls

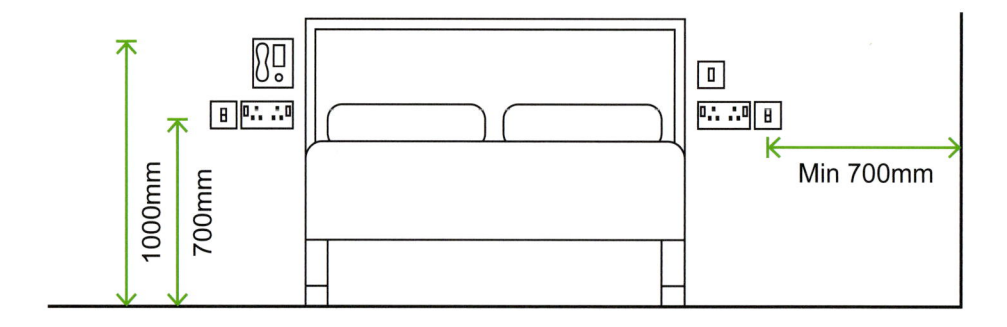

1000mm
700mm
Min 700mm

3.44b
3.44j

106   Wheelchair Housing Design Guide

# Technical provisions

> **Additional good practice recommendations**
>
> - Low surface temperature radiators should be provided in WC/cloakroom and bathroom facilities.
> - As a minimum, provide socket outlets in accordance with Table K.
>
> | Table K | Minimum number of electrical sockets within the dwelling ||
> |---|---|---|
> | **Room** | **Outlets** | **Notes** |
> | Hallway(s) | 4 | Provide a double socket in the wheelchair storage and transfer space, rather than the power socket described in 12.2.1 |
> | Liftway (where applicable) | 1 | Provide a power socket suitable for powering the lifting device close to the liftway as described in 12.2.1 |
> | Living room | 8 | Include sockets with integrated USB charging ports |
> | Dining room | 4 | |
> | Stairs | 1 | Provide a power socket suitable for powering a stair-lift close to the foot or head of the stair as described in 12.2.1 |
> | Kitchen | 8 | Include sockets with integrated USB charging ports. Additional sockets will be required for white goods |
> | Single bedroom | 6 | Include sockets with integrated USB charging ports |
> | Double bedroom | 8 | Include sockets with integrated USB charging ports |
>
> Note: Where rooms are combined, the total number of outlets should be provided. A double socket is counted as two outlets.

### 12.2.2 Consumer units

Consumer units should be mounted so that the controls are installed between 1350mm and 1450mm above floor level. — 3.44a

### 12.2.3 Door entry controls

Any door entry controls should be mounted 900mm to 1000mm above finished ground level and a minimum of 300mm away from any external return corner. — 3.22k

A fused spur suitable for the fitting of a powered door opener should be located on the hinge side of the principal private entrance door of a dwelling. — 3.22l

### 12.2.4 Doors and windows

Refer to Chapter 11: Operating internal doors and windows.

Wheelchair Housing Design Guide

# Using outdoor spaces 13

## Principle

Access to outdoor space can enhance a wheelchair user's sense of wellbeing and quality of life. Both communal and private outdoor spaces should be inclusively designed to enable ease of use and enjoyment by wheelchair users and their family and friends.

## 13.1 Design considerations

**Communal outdoor spaces**

All areas of communal gardens or landscaped areas generally available for resident use should be accessible to wheelchair users. Suitable doors, gates and paths to and from and within these spaces should be designed for wheelchair access and use.

Where communal play areas are provided, some accessible equipment should be specified to enable all children to play together. Consider providing inclusive playground equipment, such as wheelchair accessible roundabouts, nest swings or sensory panels. The surface finish in any play areas should be suitable for wheelchair access and should not be of a loose material, such as play bark chippings.

Seating should be designed to provide space alongside to enable wheelchair users to sit with a companion.

Where the outdoor space is a balcony, access deck or terrace, there should be an accessible threshold between the internal and external surfaces. Any drainage channels or grates should be designed to be flush with the surrounding surface and not have wide slots or holes, which can be a hazard to wheelchair users. There should be adequate space, clear of door swings, for a wheelchair user to use all the available space. It should also be possible to see through any balustrade or guarding[6] from a seated position.

### Figure 13.1 Balcony balustrade or guarding

Drainage channels flush with surrounding surface

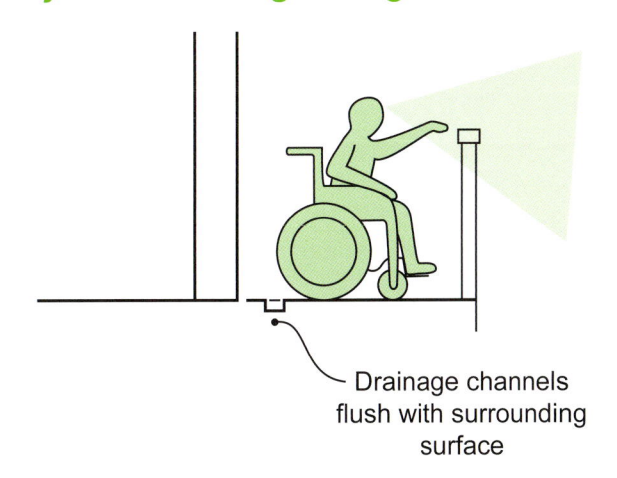

---
[6] Refer to Approved Document K.

# 13 Using outdoor spaces

Raised flower beds and planters can provide accessible gardening opportunities for wheelchair users. Where provided, they should be in a suitable location with wheelchair access.

**Private outdoor spaces**

Some people may feel vulnerable living in a ground floor dwelling where their windows open onto public paths or communal spaces. Front gardens are the preferred method of defining this boundary between public and private spaces. Where these are not provided, defensible space should be established by low walls, fences and robust low maintenance planting.

There should be step-free access through any door, from inside the dwelling to private external spaces, such as a garden or balcony. All external doors should have an accessible threshold between the internal and external surfaces. Rear external spaces and gardens should be securely enclosed and overlooked from within the dwelling. External gates should be capable of being locked and unlocked from each side so a wheelchair user is able to enter and leave securely via this route if preferred.

The layout of a garden should enable a wheelchair user to access and enjoy all areas. It can be difficult to push a wheelchair over grass, and gardens should ideally have a level patio area that is large enough to be accessed by a wheelchair user. Paths, with adequate space to turn, should be provided to facilities such as clothes drying, storage and refuse areas and any alternative entrances. Where outdoor clothes-drying facilities are provided, they should be easily accessible with a simple means of adjusting the height of the washing lines from a seated position.

### Figure 13.2   Private outdoor space

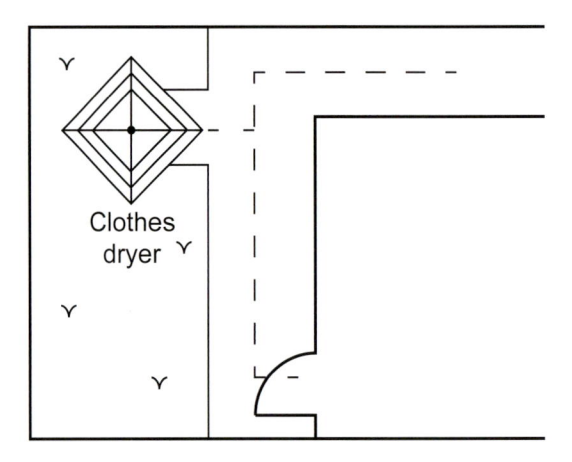

Design considerations for private balconies and terraces are the same as those for communal balconies and terraces.

# Technical provisions

## 13.2 Technical provisions

*When designing to Part M of the Building Regulations, Approved Document M, Volume 1, M4(3) Category 3, the technical provisions detailed in this chapter are relevant to both M4(3)(2)(a) – **Wheelchair adaptable dwellings** and M4(3)(2)(b) – **Wheelchair accessible dwellings**.*

### 13.2.1 Communal outdoor space

Doors or gates into communal gardens or onto terraces, access decks or balconies should meet the requirements of 3.2.3 Communal doors and gates.

Perf – a
3.4
3.15

> **Additional good practice recommendations**
>
> - All areas within communal gardens or landscaped areas available for resident use should be accessible to wheelchair users. Paths within these areas should be a minimum of 1200mm wide. A 1500mm turning circle should be provided at the end of paths and be available on communal terraces, access decks and balconies.
>
> - Where play areas are provided, chose equipment that is accessible to children with a variety of access requirements.
>
> - All paths within communal gardens and landscaped areas should have a firm, even surface that is slip resistant with no loose materials. There should be no upstands where there is a change of material, e.g. between a path and a grassed area.

### 13.2.2 Private outdoor space and storage areas

A wheelchair user should be able to access and use every private outdoor space that is provided, whether a private garden, balcony or roof terrace. All private outdoor space should have a minimum clear width of 1500mm. There should also be a minimum 1500mm clear turning circle with a level surface, no steeper than 1:60. This space should be free of any door swing.

3.19
3.45
3.45a

All outdoor paved areas should have suitable ground surfaces. These should be firm, even, smooth enough to be wheeled over, and should not be covered in loose laid materials such as gravel. Paved areas should have a maximum crossfall of 1:40.

3.45f

All paths, including those providing access to private refuse, recycling, cycle, scooter or other external stores, should have a minimum clear width of 1050mm. They should be level or gently sloping with sufficient space for a clear turning circle, of minimum diameter 1500mm, at the end of each path.

3.45b
3.45c

Every gate or gateway leading to, or associated with, private outdoor spaces should have a minimum clear opening width of 850mm, with a minimum 300mm nib to the leading edge (pull side) and a minimum 200mm nib to the following edge (push side).

3.23
3.45d
3.45e

All doors associated with, or providing access to, private outdoor spaces as well as any doors to private storage associated with the dwelling, should meet the requirements detailed in 4.2.2 Other private external doors.

3.23

# 13 Technical provisions

### Figure 13.3  Doors to balconies and terraces

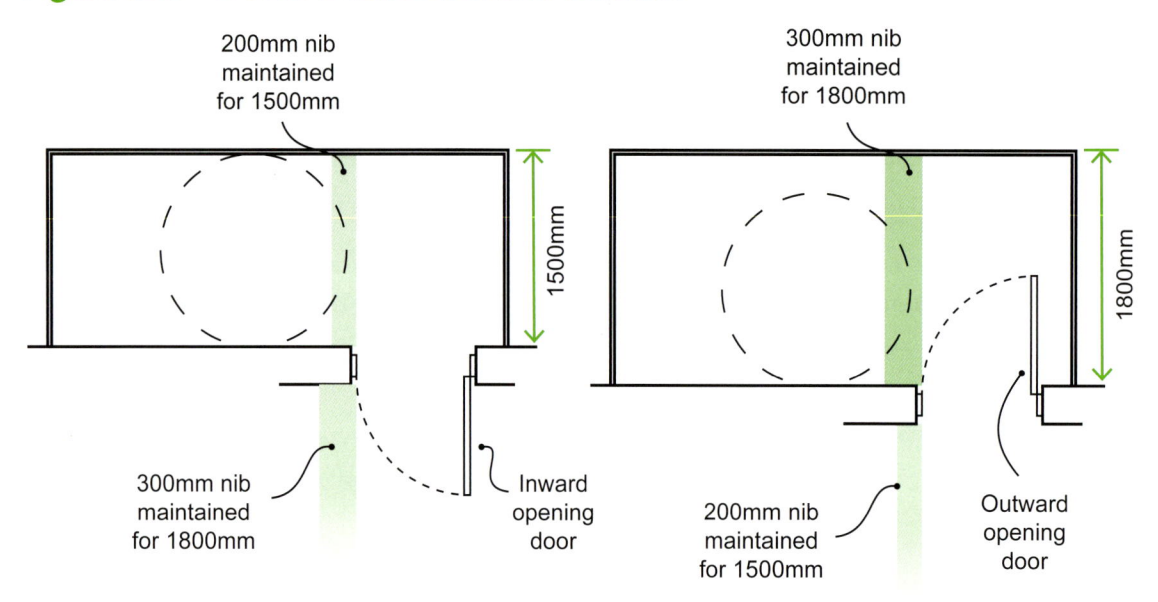

AD M
3.45a
3.23

Note: Where doors to balconies or terraces are outward opening or sliding doors, the extra width created by the 300mm nib to the leading edge of the door should be maintained for a distance of 1800mm beyond it, affecting the overall size of the balcony. Also refer to 4.2.2 Other private external doors

> **Additional good practice recommendations**
> - Allow a minimum clear width of 1200mm for all private outdoor paths.
> - Provide a suitable path to any washing line (where provided).
> - Provide fully diffused luminaires at all external private entrances, including balconies, terraces and any storage facilities associated with the dwelling.

Wheelchair Housing Design Guide

# Designing wheelchair adaptable dwellings 14

## Principle

*When designing to Part M of the Building Regulations, Approved Document M, Volume 1, M4(3) Category 3, the technical provisions detailed in this chapter are **only** relevant to M4(3)(2)(a) – **Wheelchair adaptable dwellings**.*

M4(3)(2)(a) – Wheelchair adaptable dwellings may require some simple adaptations to be suitable for occupation by a wheelchair user, while M4(3)(2)(b) – Wheelchair accessible dwellings are designed and constructed to be suitable for a wheelchair user to live in on completion.

This chapter focuses on M4(3)(2)(a) – Wheelchair adaptable dwellings and the differences between these dwellings and M4(3)(2)(b) – Wheelchair accessible dwellings. It is important to note that the majority of technical provisions for wheelchair accessible dwellings are also applicable to dwellings designed to be wheelchair adaptable. There are no differences in provision, except those set out in this chapter. The provisions within this chapter should therefore be read in conjunction with Chapters 1 to 13.

When designing an M4(3)(2)(a) – Wheelchair adaptable dwelling, it should be shown that the dwelling is capable of easy adaptation to meet the full requirements of M4(3)(2)(b) – Wheelchair accessible dwellings, without having to alter structure or services. This can be demonstrated by providing plans which show both a wheelchair adaptable and wheelchair accessible layout. A sensible approach is therefore to design the wheelchair accessible dwelling first, removing or altering the features outlined in 14.2.3 to 14.2.7 to achieve the desired plan for a wheelchair adaptable dwelling.

## 14.1 Design considerations

A home that is designed to be wheelchair adaptable will have the potential to provide the same level of accessibility as a home of the equivalent number of bedspaces designed to be wheelchair accessible from the outset. Simple adaptations should be possible without compromising required space, access, minimum storage and general usability within the home and adaptation should not result in loss of bedspaces.

Within the dwelling, the layout of a wheelchair adaptable home will be largely the same as the potential wheelchair accessible layout. This is to enable future adaptations to be made with minimal costs and disruption.

A wheelchair adaptable layout can only differ from the potential wheelchair accessible layout in the five areas detailed below.

# 14 Designing wheelchair adaptable dwellings

### Wheelchair storage and transfer space

As for wheelchair accessible dwellings, a dedicated space for wheelchair storage and transfer should be provided in wheelchair adaptable dwellings. In a wheelchair adaptable home, however, this identified space can be used for other purposes, such as other storage, with easily removable cupboard doors if desired. Most wheelchair users will have more than one wheelchair and, on occupation, can remove any cupboard doors to this area to enable easy access, storage and transfer between wheelchairs. It is important that this storage area is in addition to any minimum requirements for circulation space, living areas or general built-in storage so that stored wheelchairs do not become an obstacle or trip hazard.

### Lifting device provision

Where a wheelchair adaptable dwelling has more than one storey, a space must be identified for a lift; however, an installed lift is not required from the outset. Most wheelchair users will require the installation of a vertical lift on occupation to enable them to move between storeys while seated in their wheelchair. This identified space should be located so that it does not compromise space or circulation anywhere else in the dwelling.

A stair-lift may be an alternative adaptation; however, it does not replace the requirement for space for the liftway at design stage, as a stair-lift is likely to be unsuitable for most wheelchair users. (A stair-lift is typically used by ambulant disabled people although may be appropriate for some intermittent or occasional wheelchair users.)

### Kitchen

A wheelchair adaptable kitchen does not need to be fully fitted out to be useable by a wheelchair user from the outset. It should, however, be large enough to accommodate the additional worktop length required for a wheelchair accessible kitchen.

As a minimum, wheelchair adaptable dwellings should provide sufficient circulation space to move around the kitchen, access the dining table and have enough kitchen worktop for food preparation and worktop appliances.

A wheelchair user's ability and need to use the kitchen will influence the level of adaptation necessary in the kitchen on occupation. Adaptations for a wheelchair user potentially include the fitting of pull down/pull out accessible storage solutions inside units, replacing the oven with one at a wheelchair accessible height and installing a section of height adjustable worktop housing the hob and sink. In order for this height adjustable worktop to be wheelchair accessible, base units previously underneath it should be removed to create clear leg room. There should be sufficient space within a wheelchair adaptable layout to lengthen the worktop and provide additional units to replace this lost storage.

Where practicable, considerations for the overall layout of a wheelchair accessible kitchen, with specific thought given to the positioning of services and drainage (as outlined in Chapter 8), should be incorporated in wheelchair adaptable designs from the outset to minimise the cost and disruption of any future adaptation.

# Designing wheelchair adaptable dwellings — 14

### Bathrooms

There is some flexibility in bathroom design for wheelchair adaptable dwellings; however, it is important to ensure that the bathroom is large enough, and designed in such a way, that the access zones and other provisions of a wheelchair accessible bathroom are easily achievable with minimal disruption. A wheelchair adaptable home may have a bath fitted over drainage suitable for a future level access shower and a typical adaptation for a wheelchair user will involve removal of the bath.

Where a wheelchair adaptable bathroom has been built with the basin located adjacent to the WC, this may need to be relocated to provide clear space alongside the WC for a wheelchair user to complete a side transfer. Positioning of waste and drainage within a wheelchair adaptable bathroom should be carefully considered at the planning stage to ensure the required access zones for a wheelchair accessible bathroom are achievable without the encroachment of stacks, pipes or boxing-in.

Design considerations for the overall layout of an accessible bathroom (as outlined in Chapter 10) should be factored into wheelchair adaptable dwellings from the outset to avoid unnecessary adaptations.

### Services and controls

There is no difference between a wheelchair adaptable dwelling and a wheelchair accessible dwelling in terms of services and controls, with the exception of the requirements for the location of isolators to kitchen appliances and height requirements for radiator controls.

# 14 Technical provisions

## 14.2 Technical provisions

*When designing to Part M of the Building Regulations, Approved Document M, Volume 1, M4(3) Category 3, the technical provisions within this chapter are **only** relevant to M4(3)(2)(a) – **Wheelchair adaptable dwellings**. When designing a wheelchair adaptable dwelling, it is necessary to show that the requirements for an M4(3)(2)(b) – Wheelchair accessible dwelling (of the same number of bedspaces) can also be met and the provisions within this chapter should therefore be read in conjunction with Chapters 1 to 13.*

To demonstrate that a dwelling is capable of meeting the functional and spatial provisions for a wheelchair adaptable dwelling, furnished plan layouts that show the necessary access zones, other provisions and the furniture of the furniture schedule included in Appendix 1, should be shown at a scale of at least 1:100. Plans should also show how the full requirements of a wheelchair accessible dwelling would be met.

> **Additional good practice recommendations**
> 
> - Plans are required for Building Regulations approval; however, it is recommended that fully furnished layouts are designed at planning stage to ensure that an accessible and functional layout can be achieved.

### 14.2.1 Approach to the dwelling and communal areas and facilities

There are no differences in requirements between wheelchair adaptable and wheelchair accessible dwellings with regard to the approach route to the dwelling and communal areas or facilities. The technical provisions in Chapters 1 to 3 are therefore equally applicable for wheelchair adaptable and wheelchair accessible dwellings.

### 14.2.2 Private entrances and spaces within, and connected to, the dwelling

The only differences in requirements between wheelchair adaptable and wheelchair accessible dwellings with regard to private entrances and spaces within, or connected to, the dwelling, have been detailed in this chapter. The technical provisions in Chapters 4 to 13 are therefore equally applicable to wheelchair adaptable and wheelchair accessible dwellings, with the only exceptions outlined in 14.2.3 to 14.2.7.

Note 1: There is no difference in furniture requirements or minimum floor areas for bedrooms in a wheelchair adaptable or wheelchair accessible dwelling. Plans should clearly evidence how the proposed adaptation will meet the requirements of an equivalent wheelchair accessible dwelling without compromising these minimum bedroom requirements or reducing the number of bedspaces. Other provisions within the dwelling, such as minimum floor area for living spaces, minimum door and corridor widths and minimum number of sanitary facilities, also remain unaffected.

Note 2: In dwellings of up to four bedspaces, there should be no difference in overall dwelling size between the wheelchair adaptable dwelling and the proposed wheelchair accessible dwelling. However, in dwellings of five bedspaces or more, different requirements for sanitary facilities may result in a wheelchair adaptable dwelling that is slightly smaller overall than the wheelchair accessible equivalent (refer to 14.2.6).

**AD M**

3.20
Perf – d

3.6

3.21

Perf – d
3.6
3.21

3.43c
3.42b

# Technical provisions 14

## AD M

### 14.2.3 Wheelchair storage and transfer space (also refer to Chapter 5)

The dwelling should have a suitable storage and transfer space to enable a wheelchair user to charge and store up to two wheelchairs and transfer between an outdoor and an indoor wheelchair. The requirements outlined in 5.2.3 are equally relevant to wheelchair adaptable and wheelchair accessible dwellings.

3.25
3.25a
3.25b
3.25c

However, in wheelchair adaptable dwellings the wheelchair storage and transfer space may be used for another purpose, such as general storage (with doors fitted if required), provided that the requirements of 5.2.3 can be met without alteration to structure or services. The wheelchair storage and transfer space remains additional to the requirements for built-in storage, living spaces and bedrooms as set out in Table B in Chapter 5, Table C in Chapter 7 and Chapter 9: Technical provisions.

3.25d

### 14.2.4 Dwelling lift requirements (also refer to Chapter 6)

Provision should be made to ensure that a wheelchair user can access all parts of a dwelling on more than one floor level. Where a dwelling is defined as wheelchair adaptable, it should be easy to install a lift.

3.27
Perf – b
3.28

Future provision for the liftway should be a minimum of 1100mm wide and 1650mm long internally linking circulation areas at every floor level of the dwelling.

3.28b

In a wheelchair adaptable home, this potential liftway can be used for another purpose (such as storage). Any floors, walls and doors, however, that have been installed to allow the potential liftway to be used as storage, or for any other purpose, should be easily removable without the need for structural alteration.

3.28
3.28a

Where walls forming the liftway enclosure are not installed initially, they should be easy to install without the need for structural works and should not compromise compliance with any other provisions within this guide.

3.28c

The space to install a future lift should not be used to meet any other requirements and in particular should not be included in the minimum floor area for living, dining and kitchen spaces set out in Table C in Chapter 7 or the minimum area for built-in storage set out in Table B in Chapter 5.

3.28e

Drawings should demonstrate how all the provisions of 6.2.1 for wheelchair accessible dwellings can be complied with if a suitable lifting device is fitted in future.

3.28d

### 14.2.5 Wheelchair adaptable kitchens (also refer to Chapter 8)

#### Location

The kitchen, principal eating area and living areas should be located on the entrance level with step-free access between these areas. The kitchen and dining area should be within the same room or connected to each other.

3.32
3.31a
3.32a

#### Basic spatial requirements

In wheelchair adaptable kitchens, the requirement for the overall length of kitchen worktop is less than that of a wheelchair accessible kitchen of the equivalent number of bedspaces. Drawings should, however, show how the additional length of worktop required in a wheelchair accessible kitchen could be accommodated and demonstrate how the kitchen could be easily adapted to meet the wheelchair accessible provisions of 8.2.2 to 8.2.7 at a

3.33a
3.34a
3.33b

Wheelchair Housing Design Guide

# 14 Technical provisions

| | AD M |
|---|---|

future date. Such adaptations should be achievable without compromising the space in any other part of the dwelling and without needing to move structural walls, stacks or concealed drainage.

In a wheelchair adaptable kitchen, a minimum length of kitchen worktop is required in accordance with Table L. This worktop length is measured through the mid-line of the worktop, not the front or rear worktop edge. Worktop length includes the sink, hob and oven.

*3.33a*
*Diag 3.8*

Note: Full height appliances or storage units should not be included in worktop length as they reduce the worktop available for food preparation and cooking.

| Table L | Minimum length of kitchen worktop, including fittings and appliances, to be fitted at completion for a wheelchair adaptable dwelling |
|---|---|
| **Number of bedspaces*** | 2 | 3 & 4 | 5 | 6–8 |
| Minimum worktop length (mm) | 4330 | 4730 | 5630 | 6730 |

*Table 3.3*

*\* For the purposes of establishing the number of bedspaces relevant to these requirements, a bedroom at or above 8.5m² and below 12.5m² is counted as one bedspace, and equal to or greater than 12.5m² as two bedspaces.*

*3.36 Note 3*

A minimum 1500mm wide clear access zone is required in front of, and between, all kitchen units and appliances. Any radiators, boxing-in or other localised obstructions should not project into the clear access zone.

*3.32b*

## Controls and lighting

All electrical controls, including sockets, should meet the requirements of Chapter 12: Services and controls, with the exception of the location of isolators to kitchen appliances and height requirements for radiator controls.

*3.44*
*3.44c*
*3.44n*

> **Additional good practice recommendations**
>
> - In order to minimise the disruption and cost of any future adaptation required to a wheelchair adaptable kitchen, incorporate as many of the design considerations from Chapter 8: Using the kitchen into the design as possible. It is recommended that a wheelchair accessible layout is designed from the outset, to ensure that this level of accessibility can be achieved.

### 14.2.6 Wheelchair adaptable bathrooms (also refer to Chapter 10)

The requirements for sanitary facilities in wheelchair adaptable dwellings aim to allow increased flexibility in layout. Drawings should, however, demonstrate how the sanitary facilities could be easily adapted to meet the wheelchair accessible provisions of 10.2.1 to 10.2.16 at a future date (with the exception noted for dwellings of five or more bedspaces below). The provisions for wheelchair adaptable sanitary facilities are described below.

*3.38b*
*3.42b*

## Access zones

The clear access zones associated with each sanitary fitting (WC, basin, bath and shower) in wheelchair adaptable dwellings, as shown in Figure 14.1.1 to 14.1.5, are smaller than those of wheelchair accessible dwellings.

*3.42a*
*3.38a*
*Diag 3.10*

# Technical provisions

## 14

AD M
Diag 3.10

**Figure 14.1.1  WC access zone (wheelchair adaptable dwellings)**

- 1100mm
- 450–500mm wall to centre line of WC
- Clear space 1000mm long, 1000mm high and 100mm deep for fixing of grab rails
- 350mm
- 1000mm

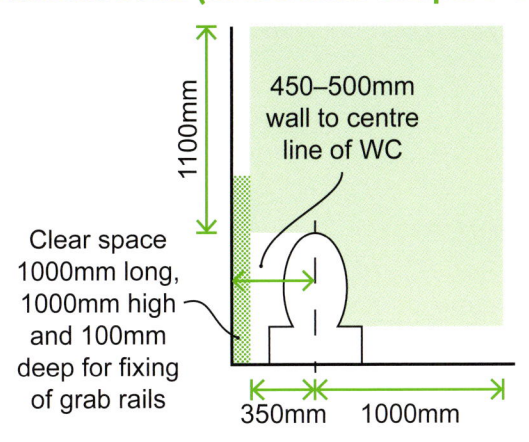

Diag 3.10

**Figure 14.1.2  Permitted basin encroachment into WC access zone (wheelchair adaptable dwellings)**

- Minimum of 750mm from edge of access zone to edge of basin
- Minimum of 750mm from front of pan to mid line of basin
- 200mm maximum permitted basin encroachment permitted anywhere inside the dotted lines
- Max 200mm
- Max 200mm
- Max 200mm
- 450–500mm

(a) Maximum permitted encroachment of 200mm for hand rinse basin (WC/cloakroom)

(b) Maximum permitted encroachment of 300mm for wash hand basin

Wheelchair Housing Design Guide

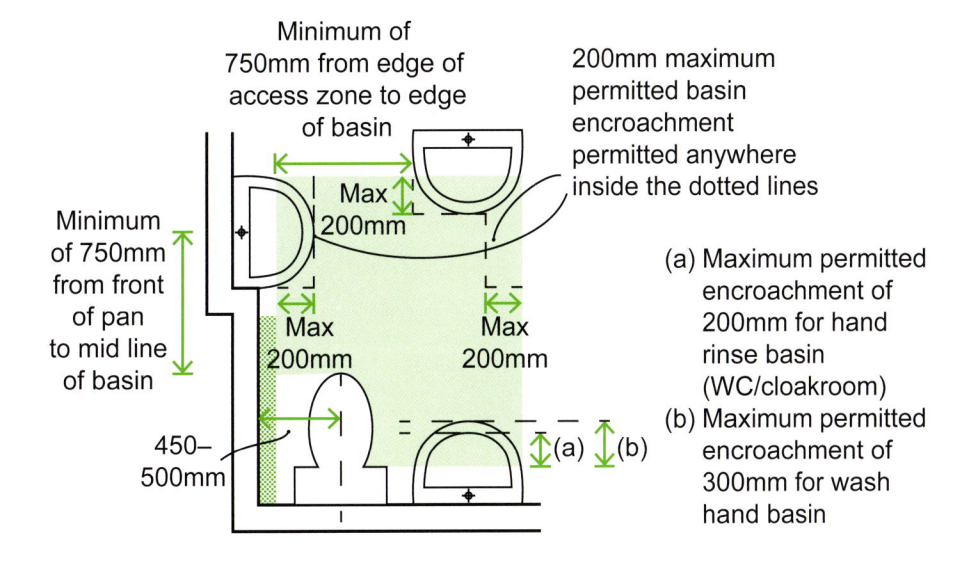

# 14 Technical provisions

### Figure 14.1.3 Basin access zone (wheelchair adaptable dwellings)

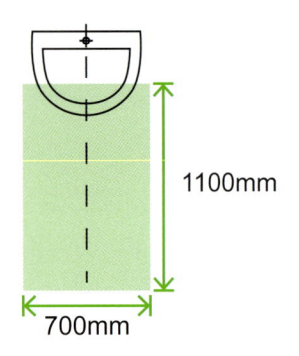

1100mm
700mm

Note: The access zone is the same for the hand rinse basin in a WC/cloakroom

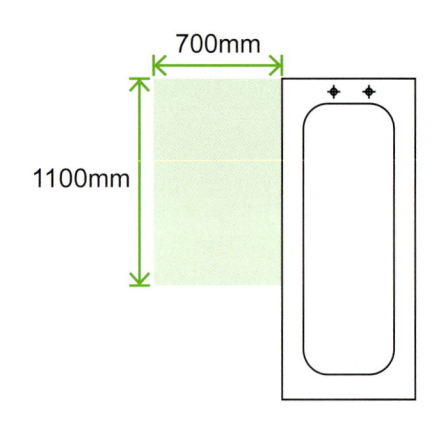

### Figure 14.1.4 Bath access zone (wheelchair adaptable dwellings)

700mm
1100mm

AD M
Diag 3.10

### Figure 14.1.5 Shower access zone (wheelchair adaptable dwellings)

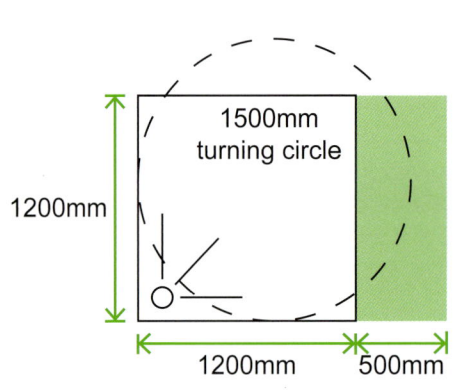

1500mm turning circle
1200mm
1200mm   500mm

Note: 1000mm × 1000mm shower permitted in WC/cloakroom

Diag 3.10

Note 1: It should be demonstrated how the wheelchair adaptable bathroom layout can be easily adapted in future to meet the wheelchair accessible clear access zones described in 10.2.4. There will therefore typically be no difference in overall size between a wheelchair adaptable layout and the proposed wheelchair accessible layout.

3.38b
3.42b
Diag 3.12
Diag 3.15
Diag 3.16

Note 2: A wash hand basin or hand rinse basin is permitted to encroach into the access zone adjacent to the WC in a wheelchair adaptable bathroom. However, drawings should show how the bathroom could be easily adapted to meet the wheelchair accessible clear access zones described in 10.2.4 with provision for an alternative basin position.

Diag 3.10
Diag 3.11

Note 3: The clear access zones for a wheelchair adaptable bathroom appear to provide greater flexibility to locate stacks and drainage. However, to ensure a future wheelchair accessible layout is achievable, the associated clear access zones are still required. Stack and drainage positions therefore need to be considered from the outset to ensure that all clear access zones required for wheelchair accessible bathroom layouts can be achieved. Bathrooms may need to be larger than the minimum sizes shown to accommodate stack and drainage positions.

Diag 3.10
Diag 3.11
3.36i

# Technical provisions 14

AD M

> **Additional good practice recommendations**
> 
> - Design a wheelchair accessible bathroom initially to ensure that this level of accessibility can be achieved. Elements of the design can then be amended in line with the relevant requirements to achieve the desired wheelchair adaptable layout.

### Minimum requirements for sanitary provision

Wheelchair accessible dwellings of five or more bedspaces require a bathroom(s) with a useable bath in addition to an installed level access shower from the outset, as outlined in 10.2.6. In wheelchair adaptable dwellings there is no requirement to show that both a useable bath and level access shower could be accommodated through future adaptation. A wheelchair adaptable dwelling with five or more bedspaces may therefore have a slightly smaller bathroom footprint than the proposed wheelchair accessible layout.

Table 3.5
3.43c
3.42b

In a wheelchair adaptable dwelling, sanitary facilities should be provided in accordance with Table M and Table N.[7]

| Table M | Summary of minimum requirements for sanitary provision in a single storey wheelchair adaptable dwelling based on number of bedspaces |  |  |  |
|---|---|---|---|---|
| **Single storey dwellings** (typically a flat or bungalow) | | **Number of bedspaces*** | | |
| | | **2 & 3** | **4** | **5 or more** |
| **Description of sanitary facility** | | **Number of required facilities** | | |
| **Bathroom with installed level access shower:** A wet room with WC, basin and an installed level access shower. Incorporate sufficient space for a potential bath, with the relevant clear access zone (if not provided elsewhere in the dwelling), to be fitted in place of the shower | | 1 | 1 | 1 |
| **Second WC/cloakroom provision (as described in 10.2.7):** As a minimum, a WC/cloakroom with a WC and basin | | | 1 | 1 |

Table 3.5

3.37a
3.41a
3.36c

3.37b
3.40

*For the purposes of establishing the number of bedspaces relevant to these requirements, a bedroom at or above 8.5m² and below 12.5m² is counted as one bedspace, and equal to or greater than 12.5m² as two bedspaces.*

3.36 Note 3

---

[7] Approved Document M, Volume 1 (2015 edition incorporating 2016 amendments) provides a summary of minimum requirements that may cover other dwelling configurations.

Wheelchair Housing Design Guide

# 14 Technical provisions

| Table N | Summary of minimum requirements for sanitary provision in two or three storey wheelchair adaptable dwellings based on number of bedspaces | | | | AD M Table 3.5 |
|---|---|---|---|---|---|
| **Two or three storey dwellings where the principal bedroom is not on the entrance storey (typically a house or maisonette)** | | **Number of bedspaces*** | | | |
| | | **2 & 3** | **4** | **5 or more** | |
| **Description of sanitary facility** | | **Number of required facilities** | | | |
| **Entrance storey WC/cloakroom with installed level access shower:** A WC/cloakroom constructed as a wet room with a WC, basin and installed level access shower | | 1 | 1 | 1 | 3.37a 3.36c |
| **Bathroom with installed level access shower:** A wet room with WC, basin and an installed level access shower, on the same storey as the principal bedroom. Incorporate sufficient space for a potential bath, with the relevant clear access zone (if not provided elsewhere in the dwelling), to be fitted in place of the shower | | 1 | 1 | 1 | 3.41a 3.41b 3.36c |

*For the purposes of establishing the number of bedspaces relevant to these requirements, a bedroom at or above 8.5m² and below 12.5m² is counted as one bedspace and a bedroom equal to or greater than 12.5m² is counted as two bedspaces.*   3.36 Note 3

Note 1: Refer to 10.2.7 for the access zones relevant to the second WC/cloakroom provision described in Table M. Remaining sanitary facilities described in Table M and Table N should incorporate the access zones of Figures 14.1.1 to 14.1.5, clear of all radiators, towel rails and services.   3.40 / 3.38a / Diag 3.10

Note 2: In wheelchair adaptable bathrooms the door may open inwards provided that the door can be easily rehung to open outwards (e.g. door stops are planted and easily moved). The door to a WC facility on the entrance storey should open outwards from the outset.   Diag 3.12 / Diag 3.15 / 3.37c

Note 3: In wheelchair adaptable dwellings, it is assumed that most commonly a bath will be fitted over the installed level access shower, although this is not a requirement.   3.41 Note 3 / 3.42

Note 4: In wheelchair adaptable dwellings of more than one storey, it would be reasonable for the entrance storey WC/cloakroom to have a potential level access shower, but it should be capable of providing an installed level access shower without the need to move walls, remove screed or other solid flooring. It should include a capped off floor gulley, set at an appropriate level, and be connected to the drainage system.   Diag 3.12

Note 5: The provisions within this chapter do not apply to sanitary facilities that are additional to the minimum provisions of Table M and Table N.   3.36 Note 2

### Minimum size of sanitary fittings

The size of all sanitary fittings should be in accordance with Table J in Chapter 10. Where larger sanitary fittings are provided, these should not reduce the clear access zone required in Figures 14.1.1 to 14.1.5.   Appx D / Diag 3.10

# Technical provisions 14

| | AD M |
|---|---|

### Services

Where located in sanitary facilities, stacks, vents, drainage and boxing-in should be clear of access zones.

*Diag 3.10*
*Diag 3.12*
*Diag 3.13*
*Diag 3.14*
*Diag 3.15*

### Supports

The ceiling structure to sanitary facilities required by Table M and Table N should be strong enough to allow the fitting of an overhead hoist capable of carrying a load of 200kg.

*3.36e*

The walls, ducts and boxings to the sanitary facilities required in Table M and Table N should be strong enough to support the fitting of grab rails, shower seats and other adaptations that could impose a load of up to 1.5kN/m². Additional localised strengthening may be required if adaptations are fitted that impose higher point loads.

*3.36d*
*3.36 Note 1*

### Controls

Taps and bathroom controls should be suitable for a person with limited grip to operate single handedly.

*3.44l*

> **Additional good practice recommendations**
>
> ● In order to minimise the disruption and cost of any future adaptation required to a wheelchair adaptable bathroom, incorporate as many of the design considerations from Chapter 10: Bathrooms into the design as possible.

### 14.2.7 Services and controls (also refer to Chapter 12)

The provision for services and controls outlined in Chapter 12 applies to wheelchair adaptable dwellings, with the exception of requirements for the location of isolators to kitchen appliances and height requirements for radiator controls.

*3.44c*

*3.44n*

# Appendices

|   |   | Page |
|---|---|---|
| 1 | Furniture schedule | 125 |
| 2 | Glossary of terms | 127 |
| 3 | Legislation and technical standards | 129 |
| 4 | Organisations and sources of information | 129 |
| 5 | Cross-referencing | 130 |

# Appendix 1: Furniture schedule

## Minimum furniture requirements for living and dining space

Also see Table D in Chapter 7 and relevant access zone requirements.

### Furniture requirements for living and dining spaces

| Furniture to be shown | Furniture size (mm) | Number of bedspaces* |||||||
|---|---|---|---|---|---|---|---|---|
| | | 2 | 3 | 4 | 5 | 6 | 7 | 8[8] |
| | | Number of furniture items required |||||||
| Arm chair (or number of sofa seats in addition to minimum sofa provision) | 850 × 850 | 2 | 3 | 1 | 2 | 3 | 4 | 5 |
| 2 seat settee (optional) | 850 × 1300 | | | | | | | |
| 3 seat settee | 850 × 1850 | | | 1 | 1 | 1 | 1 | 1 |
| TV | 220 × 650 | 1 | 1 | 1 | 1 | 1 | 1 | 1 |
| Storage units | 500 × length shown (1 only required) | 1000 | 1000 | 1500 | 2000 | 2000 | 2000 | 2000 |
| Dining table | 800 × length shown (1 only required) | 800 | 1000 | 1200 | 1350 | 1500 | 1650 | 1800 |
| Dining chair | | 2 | 3 | 4 | 5 | 6 | 7 | 8 |

\* For the purposes of establishing the number of bedspaces relevant to these requirements, a bedroom at or above 8.5m² and below 12.5m² is counted as one bedspace and equal to or greater than 12.5m² is counted as two bedspaces.

AD M

Appx D

3.36 Note 3

---

[8] Furniture requirement for eight bedspace dwellings is assumed based on Approved Document M, Volume 1 (2015 edition incorporating 2016 amendments) requirements for other dwelling sizes.

Wheelchair Housing Design Guide

# Appendix 1: Furniture schedule

## Minimum furniture requirements for bedroom

Also see Table F in Chapter 9 and relevant access zone requirements.

| Furniture requirements applicable to the principal double bedroom, other double and twin bedrooms | | |
|---|---|---|
| **Furniture to be shown** | **Furniture size (mm)** | **Quantity** |
| Principal bedroom double bed (minimum provision) | 2000 × 1500 | 1 |
| Other double bedroom double bed; or | 1900 × 1350 | 1; or |
| Twin beds (2 single beds) | 1900 × 900 | 2 |
| Bedside table | 400 × 400 | 2 |
| Double wardrobe | 600 × 1200 | 1 |
| Chest of drawers | 450 × 750 | 1 |
| Desk | 500 × 1050 | 1 |
| **Furniture requirements applicable to each single bedroom** | | |
| **Furniture to be shown** | **Furniture size (mm)** | **Quantity** |
| Single bed | 1900 × 900 | 1 |
| Bedside table | 400 × 400 | 1 |
| Double wardrobe | 600 × 1200 | 1 |
| Chest of drawers | 450 × 750 | 1 |
| Desk | 500 × 1050 | 1 |

## Minimum size of sanitary fittings

Also see Table J in Chapter 10 and relevant access zone requirements

| Bathroom sanitary fittings | | WC/cloakroom sanitary fittings | |
|---|---|---|---|
| **Sanitary fitting** | **Size (mm)** | **Sanitary fitting** | **Size (mm)** |
| WC (pan and cistern) | 500 × 700 | WC (pan and cistern) | 500 × 700 |
| Wash hand basin | 600 × 450 | Hand rinse basin | 350 × 200 |
| Level access shower | 1200 × 1200 | Level access shower | 1000 × 1000 |
| Bath | 700 × 1700 | | |

# Appendix 2: Glossary of terms[9]

**Accessible threshold**

A threshold that is level or, if raised, has a total height of not more than 15mm, a minimum number of upstands and slopes and with any upstand higher than 5mm chamfered.

**Approach route**

Internal or external path or corridor usually leading to the principal private entrance of a dwelling from a defined starting point (typically the pavement immediately outside the curtilage or plot boundary).

**Bedspace**

A suitable sleeping area for one person. (A single bedroom provides one bedspace and a double or twin bedroom provides two bedspaces where these rooms also meet any other requirements.)

**Clear access route**

Clear unobstructed 'pathway' to access a window or other feature. Localised obstructions are not permitted unless specifically stated.

**Clear access zone**

Clear, unobstructed space for access or manoeuvring. Localised obstructions are not permitted unless specifically stated.

**Clear opening width**

Clear distance measured between the inside face of the door frame (or door stop) and the face of the door when open at 90 degrees. Door furniture and ironmongery may be disregarded when measuring the clear opening width.

**Clear turning circle**

Clear floor space, represented by a circle, or an ellipse, that allows a wheelchair user to turn independently in a single movement. A door swing is permitted within a clear turning space unless stated otherwise.

**Clear width**

Clear distance measured between walls or other fixed obstructions (except permitted localised obstructions) or across a path. Skirtings totalling up to 50mm total thickness and shallow projecting ducts or casings above 1800mm may be discounted when measuring clear width.

**Communal** (area, facilities or entrances)

Shared area accessed by, or intended for use of, more than one dwelling.

**Entrance storey**

The floor level (of the dwelling) on which the principal private entrance is located.

**Following edge** (of door)

The surface of a door which follows into (or faces away from) the room or space into which the door is being opened – sometimes referred to as 'the push side'.

**Gently sloping**

Gradient between 1:60 and 1:20.

**Going** (stairs)

The depth from front to back of a tread, less any overlap with the tread above.

**Installed level access shower**

Step-free area with no lips or upstands, suitable for showering, with a floor laid to shallow falls towards a floor gulley connected to the drainage system.

**Leading edge** (of door)

The surface of a door which leads into (or faces) the room or space into which the door is being opened – sometimes referred to as 'the pull side'.

**Level**

Gradient not exceeding 1:60.

---

[9] Definitions included in the Glossary are extracts from Approved Document M, Volume 1 (2015 edition incorporating 2016 amendments) and Approved Document K (2013 edition).

Wheelchair Housing Design Guide

# Appendix 2: Glossary of terms

**Liftway**

Vertical route linking all floors of a dwelling accommodating (or capable of accommodating) a lift or lifting platform.

**Localised obstruction**

Short, fixed element, such as a bollard, lighting column or radiator, not more than 150mm deep that may intrude into a path, route or corridor, that does not unduly restrict the passage of a wheelchair user.

**Manoeuvring space**

Clear floor space, represented by a rectangle which allows a wheelchair user to turn independently in a series of manoeuvres. A door swing is permitted within a clear manoeuvring space unless stated otherwise.

**Nosing**

The leading edge of a stair tread.

**Pitch line**

A line that connects the nosing of the treads of a stair.

**Potential level access shower**

Space capable of providing a level access shower without the need to move walls, remove screed or other solid flooring. It should include a capped-off floor gulley, set at an appropriate level and connected to the drainage system (usually provided within a wet room).

**Principal communal entrance**

The communal entrance (to the core of the building containing the dwelling) which a visitor not familiar with the building would normally expect to approach (usually the common entrance to the core of a block of flats).

**Principal private entrance**

The entrance to the individual dwelling that a visitor not familiar with the dwelling would normally approach (usually the 'front door' to a house or flat).

**Ramped**

Gradient between 1:20 and 1:12.

**Rise**

For stairs: the height between consecutive treads.

For ramps: the vertical distance between each end of the ramp flight.

**Step-free**

Route without steps but that may include a ramp or a lift suitable for a wheelchair user.

**Suitable ground surface**

External ground surface that is firm, even, smooth enough to be wheeled over, is not covered with loose laid materials, such as gravel and shingle, and has a maximum crossfall of 1:40.

**Suitable tread nosings**

Nosings that conform with one of the options shown in Approved Document K.

**Tapered steps**

A step in which the going reduces from one side to the other.

**Wet room**

WC or bathroom compartment with tanking and drainage laid to fall to a connected gulley capable of draining the floor area when used as a shower.

**Wheelchair accessible dwelling**

Category 3 dwelling constructed to be suitable for immediate occupation by a wheelchair user where the planning authority specifies that optional requirement M4(3)(2)(b) applies.

**Wheelchair adaptable dwelling**

Category 3 dwelling constructed with the potential to be adapted for occupation by a wheelchair user where optional requirement M4(3)(2)(a) applies.

**Winders**

See 'tapered steps'.

# Appendix 3: Legislation and technical standards

**The Building Regulations 2010**
(SI 2010/2214) (as amended):
Approved Document B:
Fire safety
Approved Document C:
Site preparation and resistance to contaminates and moisture
Approved Document K:
Protection from falling, collision and impact
Approved Document M:
Access to and use of buildings
Approved Document Q:
Security – Dwellings

**BS EN 81-41: 2010** Safety rules for the construction and installation of lifts. Special lifts for the transport of persons and goods. Vertical lifting platforms intended for use by persons with impaired mobility

**BS EN 81-70: 2003** Safety rules for the construction and installation of lifts. Particular applications for passenger and goods passenger lifts. Accessibility to lifts for persons including persons with disability

**BS 5900: 2012** Powered homelifts with partially enclosed carriers and no liftway enclosures

**BS 8300-1: 2018** Design of an accessible and inclusive built environment. Part 1: External environment. Code of practice

**BS 8300-2: 2018** Design of an accessible and inclusive built environment. Part 2: Buildings. Code of practice

**Equality Act 2010** (Disability) Regulations 2010 (SI 2010/2128)

# Appendix 4: Organisations and sources of information

Action on Hearing Loss
www.actiononhearingloss.org.uk/

Aspire www.aspire.org.uk/

Centre for Accessible Environments
www.cae.org.uk/

Design Council Inclusive Environments Hub
www.designcouncil.org.uk/what-we-do/inclusive-environments

Disabled Living Foundation www.dlf.org.uk/

Equality and Human Rights Commission
www.equalityhumanrights.com/en

Habinteg Housing Association
www.habinteg.org.uk/

Housing Learning and Improvement Network www.housinglin.org.uk/

Joseph Rowntree Foundation www.jrf.org.uk/

Leonard Cheshire Disability
www.leonardcheshire.org/

Ministry of Housing, Communities & Local Government https://www.gov.uk/government/organisations/ministry-of-housing-communities-and-local-government

Muscular Dystrophy UK (Trailblazers)
www.musculardystrophyuk.org/

Northern Ireland Department for Communities 'Wheelchair Housing'
www.communities-ni.gov.uk/wheelchair-housing

Royal College of Occupational Therapists
www.rcot.co.uk/

Royal College of Occupational Therapists Specialist Section in Housing (RCOTSS-Housing) www.rcot.co.uk/about-us/specialist-sections/housing-rcot-ss

Royal National Institute for the Blind (RNIB Cymru) www.rnib.org.uk/

Scottish Disability Equality Forum (SDEF) 'Inclusive Design Hub'
www.inclusivedesign.scot/resources/

Stirling University Dementia Services Development Centre
www.dementia.stir.ac.uk/design

Thomas Pocklington Trust
www.pocklington-trust.org.uk/

# Appendix 5: Cross-referencing

Appendix 5 cross-references the *Approved Document M, Volume 1: Dwellings, M4(3) Category 3: Wheelchair user dwellings* (2015 edition incorporating 2016 amendments) to the relevant technical provision in this *Wheelchair Housing Design Guide* (Third Edition) (WHDG).

| AD M | WHDG | AD M | WHDG |
| --- | --- | --- | --- |
| Performance a | 2.2.1, 3.2.2, 13.2.1 | 3.13a | 2.2.8 |
| Performance b | 6.2.1, 6.2.2, 7.2.1, 8.2.1, 14.2.4 | 3.13b | 2.2.8 |
| | | 3.13c | 2.2.8 |
| Performance c | Chapter 4–14 | 3.13d | 2.2.8 |
| Performance d | 14.2, 14.2.2 | 3.14 | 3.2.1 |
| Performance e | 12.2.1 | 3.14a | 2.2.2, 3.2.1, Figure 3.4 |
| 3.1 | Building Regulations and planning policy: p4–5 | 3.14b | 3.2.1, Figure 3.4 |
| | | 3.14c | 3.2.1 |
| 3.2 | 2.2.1 | 3.14d | 3.2.1, Figure 3.5 |
| 3.3 | 2.2.1 | 3.14e | 3.2.1, Figure 3.5, 3.2.3 |
| 3.4 | 2.2.1, 3.2.2, 3.2.3, 13.2.1 | 3.14f | 3.2.1, 3.2.3 |
| 3.5 | 2.2.1 | 3.14g | 3.2.1, Figure 3.5, 3.2.3 |
| 3.6 | 14.2.1, 14.2.2 | 3.14h | 3.2.1, Figure 3.5, 3.2.3 |
| 3.7 | 2.2.1, 3.2.2 | 3.14i | 3.2.1, Figure 3.5, 3.2.3 |
| 3.8 | 2.2.1, 2.2.3 | 3.14j | 3.2.1, Figure 3.6, 3.2.3 |
| 3.9 | 3.2.2 | 3.14k | 3.2.1, Figure 3.5, 3.2.3 |
| 3.9a | 2.2.1 | 3.14l | 3.2.1, 3.2.3 |
| 3.9b | 2.2.1, 3.2.2 | 3.14m | 3.2.1, 3.2.3 |
| 3.9c | 2.2.1, 3.2.2 | 3.14n | 3.2.1, 3.2.3 |
| 3.9d | 2.2.1, 3.2.2 | Diagram 3.2 | Figure 3.4, Figure 3.5, Figure 5.11 |
| 3.9e | 2.2.1 | | |
| 3.9f | 2.2.1 | 3.15 | 2.2.4, Figure 2.5, 3.2.3, 13.2.1 |
| 3.9g | 2.2.4, Figure 2.5 | | |
| 3.10 | 3.2.2 | 3.16a | 3.2.2, 3.2.4 |
| 3.10a | 2.2.2 | 3.16b | 3.2.4 |
| 3.10b | 2.2.2 | 3.16c | 3.2.4 |
| 3.10c | 2.2.2 | 3.16d | 3.2.4 |
| 3.10d | 2.2.2, Figure 2.4 | 3.16e | 3.2.4 |
| 3.10e | 2.2.2 | 3.16f | 3.2.4 |
| 3.10f | 2.2.2 | 3.17 | 3.2.5 |
| Diagram 3.1 | Table A, Figure 2.4 | 3.18 | Building Regulation and planning policy: p4–5 |
| 3.11 | 2.2.3 | | |
| 3.11a | 2.2.3 | 3.19 | 4.2.2, 13.2.2 |
| 3.11b | 2.2.3 | 3.20 | 7.2, 7.2.3, 8.2, 9.2, 9.2.3, 10.2, 14.2 |
| 3.11c | 2.2.3 | | |
| 3.11d | 2.2.3 | 3.21 | 14.2.2 |
| 3.11e | 2.2.3 | 3.22 | 4.2.1 |
| 3.11f | 2.2.3 | 3.22a | 2.2.2, 3.2.2, 4.2.1 |
| 3.11g | 2.2.3 | 3.22b | 4.2.1 |
| 3.12 | 2.2.5 | 3.22c | 4.2.1 |
| 3.12a | 2.2.6, Figure 2.6 | 3.22d | 4.2.1 |
| 3.12b | 2.2.7, Figure 2.7 | 3.22e | 4.2.1 |
| 3.12c | 2.2.5 | 3.22f | 4.2.1, 4.2.2 |
| 3.12d | 2.2.6, Figure 2.6, 2.2.7 | 3.22g | 4.2.1, 4.2.2 |
| 3.12e | 2.2.5 | 3.22h | 4.2.1, 4.2.2 |
| 3.12 Note | 2.2.7, Figure 2.7 | 3.22i | 4.2.1, 4.2.2 |
| 3.13 | 2.2.8 | 3.22j | 4.2.1, 4.2.2 |

# Appendix 5: Cross-referencing

| AD M | WHDG |
|---|---|
| 3.22k | 4.2.1, Figure 4.8, 4.2.2, 12.2.3 |
| 3.22l | 4.2.1, 12.2.3 |
| Diagram 3.3 | Figure 4.7, Figure 4.9 |
| 3.23 | 4.2.2, 13.2.2, Figure 13.3 |
| 3.24 | 5.2.1 |
| 3.24a | 5.2.1 |
| 3.24b | 5.2.1 |
| 3.24c | 5.2.1 |
| 3.24d | 5.2.2 |
| 3.24e | 5.2.1 |
| 3.24f | 5.2.2 |
| 3.24g | 5.2.2 |
| 3.24 Note 1 | 5.2.2 |
| 3.24 Note 2 | 5.2.2 |
| Diagram 3.4 | 5.2.1, Figure 5.9, 5.2.2 |
| Diagram 3.5 | 5.2.1, Figure 5.10 |
| 3.25 | 5.2.3, 14.2.3 |
| 3.25a | 5.2.3, 14.2.3 |
| 3.25b | 5.2.3, 14.2.3 |
| 3.25c | 5.2.3, 12.2.1, 14.2.3 |
| 3.25d | 5.2.3, 14.2.3 |
| Diagram 3.6 | Figure 5.12 |
| 3.26 | 5.2.4 |
| Table 3.1 | Table B |
| 3.26 Note | 5.2.4 |
| 3.27 | 6.2.1, 14.2.4 |
| 3.28 | 14.2.4 |
| 3.28a | 14.2.4 |
| 3.28b | 14.2.4 |
| 3.28c | 14.2.4 |
| 3.28d | 14.2.4 |
| 3.28e | 6.2.1, 14.2.4 |
| 3.29 | 6.2.1 |
| 3.29a | 6.2.1 |
| 3.29b | 6.2.1 |
| 3.29c | 6.2.1 |
| 3.29d | 6.2.1, 12.2.1 |
| 3.29e | 6.2.1 |
| 3.29f | 6.2.1 |
| 3.29g | 6.2.1 |
| 3.29 Note | 6.2.1 |
| Diagram 3.7 | 6.2.1, Figure 6.1 |
| 3.30 | 6.2.2 |
| 3.30a | 6.2.2, 7.2.1, 8.2.1 |
| 3.30b | 6.2.2 |
| 3.30c | 6.2.2, Figure 6.2 |
| 3.30d | 6.2.2, 12.2.1 |
| 3.30e | 6.2.2 |
| 3.31 | 7.2.1 |
| 3.31a | 7.2.1, 8.2.1, 14.2.5 |
| 3.31b | 7.2.2 |
| 3.31c | 7.2.4, Footnote 3 |

| AD M | WHDG |
|---|---|
| Table 3.2 | 7.2.2, Table C |
| 3.32 | 7.2.1, 8.2.1, 14.2.5 |
| 3.32a | 7.2.1, 14.2.5 |
| 3.32b | 8.2.2, 14.2.5 |
| 3.33a | 14.2.5 |
| 3.33b | 14.2.5 |
| Table 3.3 | Table L |
| Diagram 3.8 | 8.2.2, Figure 8.8, 8.2.3, Figure 8.9, Figure 8.12, Figure 8.13, 14.2.5 |
| 3.34a | 8.2.2, 14.2.5 |
| 3.34b | 8.2.3 |
| 3.34c | 8.2.4 |
| 3.34d | 8.2.4 |
| 3.34e | 8.2.6, Footnote 4 |
| 3.34f | 8.2.6 |
| 3.34g | 8.2.3 |
| 3.34h | 8.2.4 |
| 3.34i | 8.2.4 |
| Table 3.4 | Table E |
| 3.35 | 9.2.1 |
| 3.35a | 9.2.4, Table G, Figure 9.3, Figure 9.4 |
| 3.35b | 9.2.4, Table G, Figure 9.3, Figure 9.4 |
| 3.35c | 9.2.6 |
| 3.35d | 9.2.1, 9.2.2 |
| 3.35e | 9.2.4, Figure 9.2, Table G, Figure 9.3 |
| 3.35f | 9.2.2 |
| 3.35g | 9.2.4, Figure 9.2, Table G |
| 3.35h | 9.2.4, Figure 9.2, Table G, Figure 9.4 |
| 3.35i | 9.2.2 |
| 3.35 Note 1 | 9.2.3 |
| 3.35 Note 2 | 9.2.6 |
| Diagram 3.9 | 9.2.4, Figure 9.3, Figure 9.4 |
| 3.36 | 10.2.1 |
| 3.36a | 10.2.1–10.2.15, 14.2.6 |
| 3.36b | 10.2.1, 10.2.7 |
| 3.36c | 10.2.1, Table H, Table I, 10.2.5, 10.2.6, 10.2.8, Table M, Table N |
| 3.36d | 10.2.16, 14.2.6 |
| 3.36e | 10.2.16, 14.2.6 |
| 3.36f | 10.2.10 |
| 3.36g | 10.2.10 |
| 3.36h | 10.2.13, Figure 10.11 |
| 3.36i | 14.2.6 |
| 3.36 Note 1 | 10.2.16, 14.2.6 |
| 3.36 Note 2 | 10.2.2, 14.2.6 |

# Appendix 5: Cross-referencing

| AD M | WHDG |
|---|---|
| 3.36 Note 3 | Table C, Table D, Table E, Table H, Table I, Table L, Table M, Table N, Appendix 1 |
| Table 3.5 | 10.2.1, 10.2.2, Table H, Table I, 14.2.6, Table M, Table N |
| 3.37a | 10.2.1, Table H, Table I, 10.2.8, 10.2.11, Table M, Table N |
| 3.37b | Table H, 10.2.7, Table M |
| 3.37c | 10.2.1, 10.2.8, 14.2.6 |
| 3.38a | 14.2.6 |
| 3.38b | 14.2.6 |
| Diagram 3.10 | 14.2.6, Figure 14.1.1–14.1.5 |
| 3.39a | 10.2.4, 10.2.5, 10.2.6, 10.2.8, Figure 10.8.1–10.8.4 |
| Diagram 3.11 | 10.2.1, 10.2.3, Table J, 10.2.4, 10.2.5, Figure 10.3.1–10.3.5, Figure 10.4, 10.2.6, Figure 10.5, Figure 10.8.1–10.8.4, Figure 10.9, Figure 10.11, 10.2.13, 10.2.14, 14.2.6 |
| Diagram 3.12 | 10.2.1, Figure 10.9, 10.2.14, 14.2.6 |
| 3.40 | Table H, 10.2.7, Table M, 14.2.6 |
| Diagram 3.13 | 10.2.1, 10.2.14, 10.2.7, Figure 10.6.1, Figure 10.6.2, Figure 10.6.3, Figure 10.7, 10.2.14, 14.2.6 |
| Diagram 3.14 | Figure 10.7, 10.2.14, 14.2.6 |
| 3.41a | 10.2.1, Table H, Table I, 10.2.5, Table M, Table N |
| 3.41b | 6.2.2, 9.2.1, 10.2.1, Table I, 10.2.5, 10.2.6, Table N |
| 3.41 Note 1 | 10.2.6 |
| 3.41 Note 2 | 10.2.6 |
| 3.41 Note 3 | 14.2.6 |
| 3.42 | 14.2.6 |
| 3.42a | 14.2.6 |
| 3.42b | 14.2.2, 14.2.6 |
| 3.43a | 10.2.1, 10.2.3, Figure 10.3.1–10.3.5 |
| 3.43b | 10.2.5 |
| 3.43c | 10.2.1, Table H, Table I, 10.2.6, 14.2.2, 14.2.6 |
| 3.43d | 10.2.11 |

| AD M | WHDG |
|---|---|
| 3.43e | 10.2.4, 10.2.5, 10.2.6 |
| Diagram 3.15 | 10.2.1, 10.2.14, 14.2.6 |
| Diagram 3.16 | 10.2.1, Figure 10.4, 10.2.14, 14.2.6 |
| Diagram 3.17 | 10.2.1, Figure 10.5, 10.2.14 |
| 3.44 | 12.2.1, 14.2.5 |
| 3.44a | 12.2.2 |
| 3.44b | 5.2.3, 7.2.5, 9.2.5, Figure 9.5, 12.2.1, Figure 12.2, Figure 12.4 |
| 3.44c | 12.2.1, 14.2.5, 14.2.7 |
| 3.44d | 7.2.4, Figure 7.5, 11.2.2, Figure 11.3 |
| 3.44e | 7.2.4, 11.2.2 |
| 3.44f | 3.2.1, 3.2.3, 4.2.1, 4.2.2, 11.2.1, Figure 11.2 |
| 3.44g | 12.2.1, Figure 12.3 |
| 3.44h | 12.2.1, Figure 12.3 |
| 3.44i | 3.2.1, 4.2.1, 7.2.5, 9.2.5, Figure 9.5, 12.2.1 |
| 3.44j | 9.2.5, Figure 9.5, 12.2.1, Figure 12.4 |
| 3.44k | 12.2.1 |
| 3.44l | 10.2.11, 10.2.12, 10.2.13, 14.2.6 |
| 3.44m | 12.2.1 |
| 3.44n | 12.2.1, 14.2.5, 14.2.7 |
| 3.45 | 13.2.2 |
| 3.45a | 13.2.2, Figure 13.3 |
| 3.45b | 13.2.2 |
| 3.45c | 13.2.2 |
| 3.45d | 13.2.2 |
| 3.45e | 13.2.2 |
| 3.45f | 13.2.2 |
| Appendix A | Appendix 2: Glossary |
| Appendix B | Appendix 3: Technical Standards |
| Appendix C | Appendix 3: Technical Standards (where applicable) |
| Appendix D | 5.2.4, 7.2.3, Table D, Table F, Figure 9.1, 10.2.3, Table J, 10.2.10, 10.2.12, 14.2.6, Appendix 1 |

# Index

Note: page numbers in *italics* refer to figures; page numbers in **bold** refer to tables.

## A

access route to windows  56, 75, **76**, *76, 77*, 77
access zones  127
    bathroom  84, 87, *88–91, 92, 93–4*, 96, *97*, 119, *119-120*, 120
    bedrooms  75–7, **76**, *76, 77*, **77**
    kitchen  65, 118
    parking  16, *16*
accessible thresholds  19, 25, *25*, 26, 32, 34, 36, 39, *41*, 109, 110, 127
adaptations to existing homes  3
approach routes  9–12, 13-15, 19-22, 23-27, 109, 111, 116, 127
    doors and gates on  10, 15, 26, 111
        nibs and reveals  10, 15, 24, 26
        opening width  15, 24, 26
    landings  9, 10, 14, 15
automatic toilets  80

## B

balconies  109, 110
    doors onto  30, 32, 36, 111, 112, *112*
balustrades  109
basins  82-3, **87**, 96-7, **126**
bathrooms  79–98, 115, 118-123
    accessories  83
    access from bedroom  79, 95
    access zones  84, 87, *88–91, 92, 93–4*, 96, *97*, 119, *119-120*, 120
    door opening  80, 84, 122
    door unlocking  101
    floor and wall finishes  83, 97
    grab rails  82, 83, *88, 93, 96,* 98, *119,* 123
    heating  83 (*also see* radiators)
    minimum size of sanitary fittings  86, **87**, 122, **126**
    services  79, 87, *90, 91, 92, 94*, 97, 123
    taps  82, 96, 123
    wall and ceiling structure  83, 98, 123
    wheelchair adaptable dwellings  115, 118–23
baths  82, **87**, 96, **126**
bedrooms  71–8
    bedhead controls  72, 77–8, 104, 106
    access zones  75–7, **76**, *76, 77*, 77
    furniture requirements  73, **74**, *74*, **126**
    hoists  72, 77, 78
    location  71, 73
    room sizes  73
    services and controls  103-7
    wheelchair adaptable dwellings  116
bedspaces  73, 127
bespoke design  2–3
boiler timer controls and thermostats  105
British Standards  27, 50, 129
Building Regulations  2, 4, 129
built-in storage  41, 45, **45** (*also see* storage)

## C

canopy or covered landing  *see* communal entrance covers; principal private entrance
carports  12, 16
ceiling structure  72, 78, 83, 98, 123
circulation areas  *see* communal corridors; dwelling circulation areas
clear access route  127
clear access zone  127 (*see also* access zone)
clear opening width  15, 21, 24, 26, 32, 34, 36, 39, 40, *40*, 41, 43, *43*, 111, 127
clear turning circle  *see* turning circle
clear width  127
clothes-drying  63, 110
communal, definition  127
communal corridors  21, 26
communal doors  19–21, 23–4 (*see also* approach routes, doors and gates on)
    entry controls  23, 25, 27, 56, 78
    fail-safe hold open devices  21
    ironmongery  23, 27
    nibs  19, 20, *20*, 24
    opening width  21, 24, 26
    power assisted  19, 25, 27
    thresholds  19, 25, *25*, 26
    vision panels  21, 27
communal entrance covers  19, 23, 25
communal entrance landings  19, 23, *23*
communal facilities  11, 13, 19, 26, 109-10, 111
communal gardens  11, 13, 109-10, 111
communal lifts  22, 27
communal lobbies  21, 24, 26
communal outdoor spaces  11, 13, 109–10, 111
communal parking bays  11, 12, 15, 16
communal play areas  11, 109, 111
communal staircases  27
communal storage areas  11, 12, 13

# Index

consumer units  104, 107
corridors  *see* communal corridors; dwelling corridors
crossfalls  7, 9, *9,* 13, 15, 17, 111
cycle stores  11

## D

door closers  20, 30, 100, 101
door handles and locks *see* handles and locks
door thresholds  19, 25, *25,* 26, 32, 34, 36, 39, *41,* 109, 110, 127
doors *see* communal doors; private external doors; internal doors (dwelling); principal private entrance
double doors  21, 24, 26, 32, 34, 36, 39
drainage channels and grates  109
drop-off points  12, 17
dual aspect windows  *52*
dwelling circulation areas  37–45
dwelling corridors  37–44
dwelling entrance covers  29, 33, *33,* 35 (*see also* communal entrance covers)
dwelling entrance doors *see* principal private entrance; private external doors
dwelling entrance lobbies  31, 34
dwelling external doors  *see* private external doors
dwelling lifts  47–8, 49–50, 114, 117

## E

emergency egress  8
entrance canopies or cover *see* communal entrance covers; principal private entrance
entrance doors *see* communal doors; principal private entrance;

entrance lobbies *see* communal lobbies; dwelling entrance lobbies
entrance storey, definition  127
entry controls
　communal entrance doors  23, 25, 27, 78
　entry phone  23, 34, 51, 56, 72, 78, 104, 106
　private entrance doors  31, 34, 35, 36, 56, 107
environmental controls  3, 72, 104
existing homes adaptation  3
external approach routes *see* approach routes
external doors *see* approach routes, doors and gates on; communal doors; principal private entrance; private external doors
external lighting  8, 11, 13, 19, 23, 25, 29, 34, 112
external seating  11, 109
external spaces  7, 11, 13, 109–12
external steps  14–15

## F

flooring
　bathrooms  83, 97
　communal corridors  26, 27
　communal entrances  21, 25
　dwelling circulation areas  39
　entrance doors  31, 35
　kitchens  63
　matting  21, 25, 31, 35
following edge (of door)  10, 15, 19, *20,* 24, 26, 29, *30,* 34, 36, *40,* 42, 44, 111, *112,* 127
French doors  32
furnished plan layouts  4–5
furniture requirements
　bedrooms  73, **74,** *74,* **126**
　living spaces  **54, 125**

## G

garages  12, **16**
gardens  7, 11, 13, 109–12
gates  10, 14, 15, 24, 26, 110, 111
　on approach routes  10, 13, 14, 15, 26, 111
　nibs  10, 15, 24, 26, 111
　opening width  15, 24, 26, 111
gently sloping, definition  127
glazed doors  51
going (stairs)  14, 127
grab rails  82, 83, *96,* 98, 99, *100,* 123
gradients  7, 9, 13–14, **14,** 17, 21, 26
ground floor dwellings  8
ground surfaces (external)  7, 10, 13, 15, 17, 111
guarding  55, 109

## H

hallway width  38, 39, 42, 43, 44
handles and locks  99, 101
　communal entrance doors  23, 27
　internal doors  101
　private entrance doors  30, 34, 36
　windows  52, 55, 100, 101–2
handrails  15
heating  83, 104 (*also see* radiators)
　boiler timer controls and thermostats  105
height adjustable worktops  59, *59,* 65–6, 114
hoists  51, 72, 78, 83, 98, 123
hot water temperature  80

## I

information sources  129
installed level access shower  79, 80–2, **87,** 95, *96,* **126,** 127
　access zones  87, *89, 94, 120*
　definition  127
　dwellings of five or more bedspaces  90–1
　dwellings up to four bedspaces  87-90

134　　　　　　　　　　　　　　　　　　　　　　　　　　　Wheelchair Housing Design Guide

# Index

entrance storey WC/cloakroom with 93-4
   minimum provision **85**, **86**
   minimum size **87**, **126**
   wheelchair adaptable dwellings **121**, **122**
internal doors (dwelling) 39–40, 43–4
   closers 101
   handles and locks 99, 101
   nibs 40, 44
   opening width 39, 40, *40*, 43, *43*
ironmongery 23, 27, 30, 32, 34, 36, 40, 99, 101

## K

kitchens 57–70
   appliances 61–3, 68
   clear access zones 65, 118
   controls 68, 118
   flooring 63
   layouts 57–8, 69–70
   lighting 63, 118
   location 57, 64, 117
   sinks 60, 67
   storage 60–1, 67
   taps 60, 67
   tiles/wall covering 63
   wheelchair adaptable dwellings 114, 117–18
   worktops 59, 64–6, 70, 114, 117–18

## L

landings
   approach routes 9, 10, 14, 15
   communal entrance doors 19, 23
   private entrance doors 29, 33
landscaping 7, 109, 111
leading edge (of door) 10, 15, 19, *20*, 24, 26, 29, *30*, 34, 36, *40*, 42, 44, 111, *112*, 127
legislation 129
letter boxes/cages 21, 25, 32, 34, *35*
level, definition 127
level access shower *see* installed level access shower
lifts *see* communal lifts; lifting platforms; liftways

lifting platforms 47–8, 49–50
liftways 47, 49-50, 114, 117, 128
light switches 29, 63, 77, 78, 83, 103, 105, 106
lighting
   communal entrances 19, 23, 25
   external lighting 8, 11, 13, 112
   kitchens 63
   natural 52, 104
   private entrance doors 29, 34, 36
living spaces 51–6
   clear access zones 53, 55, 56
   furniture requirements **54**, **125**
   room size and layout 51, 52, 53
localised obstructions 13, 24, 26, 34, *39*, 42, *42*, 65, 118, 128
low surface temperature (LST) radiators 83, 104, 107

## M

manoeuvring space, definition 128
matting 21, 25, 31, 35
mirrors 83
mobility scooter stores 7, 11, 12, 13, 20, 111
multi-storey dwellings 8, 47–50

## N

nosing 14, 128

## O

obstructions *see* localised obstructions
open plan living 51, *52*
outdoor clothes-drying 110
outdoor seating 11, 109
outdoor spaces 7, 11, 13, 109–12

## P

parking bays 11-12, 15-16
passive surveillance 31
paths 7, 110, 111, 112
   crossfalls 7, 9, 13, 111
   upstands 10, 111
   width 10, 13, 111
paving 7, 10, 13, 111
pitch line (stairs) 50, *50*, 128
planning policy 5
play areas 11, 109, 111
plumbing
   bathroom 79–80, 81, 82, 87, *90*, *91*, *92*, *94*, 95, 97
   kitchen 60, 66, 67
   wheelchair adaptable 115, 120, 122, 123
pocket doors *see* sliding doors
porches 21, 24, 26, 31, 34, 36
post boxes/cages 21, 25, 32, 34, *35*
potential level access shower 122, 128
power assisted doors 19, 25, 30, 35, 107
power sockets 103–4, 105–6
   minimum number **107**
   for powered wheelchairs 41, 44, *45*
principal communal entrance 128 (*see also* communal entrance covers; communal doors)
principal private entrance 29–36
   entry controls 31, 34, 35, 36, 56, 72, 107
   entrance cover 29, 33, *33*, 35
   flooring 31, 35
   ironmongery 30, 34
   landings 29, 33, *33*
   lobbies 31, 34
   nibs and reveals 29, *30*, 30, 34, *35*
   opening width 34, *35*
   passive surveillance 31
   power assisted 30, 35, 107

# Index

remote door release 31, 34, 51, 56, 72, 78, 104, 106
thresholds 32, 34
private external doors 29-30, 32, *32*, 36, 110–12
    double 32, 36
    entry controls 36
    ironmongery 30, 32, 36
    nibs and reveals 29, *30*, 30, 36, 111, *112*
    opening width 36, 111
    sliding 30, 36
    thresholds 32, 36, 110, 111
private outdoor spaces 110, 111–12
private outdoor storage areas 111
private stairs 48, 50
pull cord switches 83, 103, 105, *106*

## R

radiators
    in bathrooms 83, 87, 97
    in bedrooms 72, 73, *76*, **76**, 77
    in communal circulation areas 26
    controls 104, 105, 123
    in dwelling circulation areas 39, 42
    low surface temperature (LST) 83, 104, 107
raised flower beds and planters 110
ramped, definition 128
ramps 9, 13–14, **14**, *14*, 21, 26
refuse stores 11, 13, 110, 111
remote door release
    communal entrance doors 21, 23, 51, 56, 72, 78, 104, 106
    private entrance doors 31, 34, 51, 56, 72, 78, 104, 106
rise, definition 128

## S

scalding risk 67, 80, 83, 104
security 8, 29, 31, 104, 110
shower seats 81, 83, 95, *96*, 98, 123

showers 80–2, **87**, 95, *96*, **126** (*see also* installed level access shower)
single-storey homes 8
site layout 7–8
site levels 7
sliding doors 30, *30*, 36, 40, *112*
    handles and locks 40, 99, 101
    nibs *30*, 36, 40, *40*, 44, *112*
    opening width 36, 40, *40*, 43, *43*
    thresholds 32, 36
sloping *see* gently sloping; ramps
sockets *see* power sockets
specialist equipment 3, 80, 82 (*see also* hoists)
stair-lifts 48, 50, 106, 114
stairs 27, 48, 50
standards 129
step-free, definition 128
steps, external 14–15
stopcocks 104, 105
storage
    built-in 41, 45, **45**
    communal areas 11, 13
    kitchens 60–1, 67
    for mobility scooters 11, 12, 13
    private outdoor areas 111
    refuse 11, 13
    walk-in 43, 44
    for wheelchairs 41, 44, *45*, 114, 117
street furniture 11, 13
suitable ground surface 128 (*see also* ground surface)
suitable tread nosings 14, 128
switches 78, 103–4, 105–6

## T

tapered steps 14, 128
taps 60, 67, 82, 96, 123
technology 3, 104
telephone points 77, 106
terraces 11, 13, 109, 110, 111
    doors onto 32, 36, 111, 112, *112*
thermal comfort 104 (*see also* heating)
thermostats 105

thresholds *see* accessible thresholds
through-floor lifts *see* lifting platforms; liftways
travel distances 9, 19
trickle vents 100
two-way switching 77, 103, 106
turning circle 24, *24*, 34, *35*, 43, *43*, 49, *49*, 50, 53, 77, 87, 90, 93, 111, *120*

## V

vehicles 11
ventilation 51, 52, 55, 56, 100
video entry systems 25, 35, 104
views to the outside 52
vision panels 21, 27
visitable dwellings 4

## W

walk-in cupboards and storage 43, 44
wardrobes 72, **74**, *75*, **126**
wash hand basins *see* basin
washing lines 110, 112
WCs **87**, 80, 95, **126**
wet rooms 80–1, **85**, **86**, **121**, **122**, 128
wheelchair accessible dwellings, definition 128
wheelchair accessible vehicles 11
wheelchair adaptable dwellings 4, 113–23
    approach routes 116
    bathrooms 115, 118–23
    bedrooms 116
    definition 128
    kitchens 114, 117–18
    lifts 114, 117
    services and controls 115, 118, 123
    wheelchair storage and transfer space 114, 117
winders *see* tapered steps
windows
    access to 56, 71, 75-7, 100
    glazing height 52, 55, 101
    handles 100, 101–2
    height 52, 55, 101–2
    remote opening 100, 101
    ventilation 51, 52, 55, 56, 100